Maize Genetic Resources - Breeding Strategies and Recent Advances

Edited by Mohamed Ahmed El-Esawi

Published in London, United Kingdom

IntechOpen

Supporting open minds since 2005

Maize Genetic Resources - Breeding Strategies and Recent Advances
http://dx.doi.org/10.5772/intechopen.95713
Edited by Mohamed Ahmed El-Esawi

Contributors
Arfang Badji, Issa Diedhiou, Abdoulaye Fofana Fall, Denis Nsubuga, Isa Kabenge, Ahamada Zziwa, Nicholas Kiggundu, Joshua Wanyama, Noble Banadda, Meena Shekhar, Nirupma Singh, Cebisa Noxolo Nesamvuni, Khavhatondwi Rinah Netshiheni, Oluwaseun Funmi Akinmoladun, Md. Rashidul Islam, Muhtarima Jannat, Md. Mostafa Masud, Samrin Bashar, Mamuna Mahjabin Mita, Muhammad Iqbal Hossain, Md. Zahangir Alam, Sabina Yeasmin, Mushfika Nusrat, Nnadozie Okonkwo Nnoli, Ahmed Balogun, Jerome Omotosho, Samuel Agele, Ulin Antobelli Basilio-Cortes, Daniel González-Mendoza, Carlos Enrique Ail-Catzim, Carlos Ceceña-Durán, Adabella Suarez-Vargas, Onésimo Grimaldo-Juárez, Dagoberto Durán-Hernández, Olivia Tzintzun-Camacho, Ángel Manuel Suárez-Hernández, Aurelia Mendoza-Gómez, Juan Carlos Vásquez-Angulo, Luis Antonio González-Anguiano, David Cervantes-García, Gabriel Luna-Sandoval, Leonides Castellanos González, Renato de Mello Prado, Cid Naudi Silva Campos, Yousaf Ali, Taufiq Nawaz, Muhammad Junaid, Mehwish Kanwal, Fazli Hameed, Saeed Ahmed, Rafi Ullah, Nazeer Ahmed, Muhammad Shahab, Fazli Subhan

First published in London, United Kingdom, 2022 by IntechOpen
IntechOpen is the global imprint of INTECHOPEN LIMITED, registered in England and Wales, registration number: 11086078, 5 Princes Gate Court, London, SW7 2QJ, United Kingdom
Printed in Croatia

British Library Cataloguing-in-Publication Data
A catalogue record for this book is available from the British Library

Additional hard and PDF copies can be obtained from orders@intechopen.com

Maize Genetic Resources - Breeding Strategies and Recent Advances
Edited by Mohamed Ahmed El-Esawi
p. cm.
Print ISBN 978-1-80355-015-2
Online ISBN 978-1-80355-016-9
eBook (PDF) ISBN 978-1-80355-017-6

For EU product safety concerns:
IN TECH d.o.o., Prolaz Marije Krucifikse Kozulić 3, 51000 Rijeka, Croatia, info@intechopen.com or visit our website at intechopen.com.

Meet the editor

Dr. Mohamed A. El-Esawi is a visiting research fellow at the University of Cambridge, United Kingdom, and Associate Professor of Molecular Genetics, Botany Department, Faculty of Science, Tanta University, Egypt. Dr. El-Esawi received his BSc and MSc from Tanta University, and his Ph.D. degree in Plant Genetics and Molecular Biology from Dublin Institute of Technology, Technological University Dublin, Ireland. After obtaining his Ph.D., Dr. El-Esawi joined the University of Warwick, United Kingdom; University of Sorbonne, France; and University of Leuven (KU Leuven), Belgium as a visiting research fellow. His research focuses on plant genetics, genomics, molecular biology, molecular physiology, developmental biology, plant-microbe interaction, and bioinformatics. He has authored several international peer-reviewed articles, book chapters, and books, and has participated in more than sixty conferences and workshops worldwide. Dr. El-Esawi is currently involved in several biological science research projects.

Contents

Preface

Maize is one of the most economically important food crops worldwide. It is used for livestock feeds and human nutrition. Oilseed crops, such as maize, have a key role in the global agricultural economy. Recent strategies and new technologies have been developed and adopted for improving maize crop production and productivity.

This book brings together recent advances, breeding strategies, and applications that have been made and applied in the field of biological control, breeding, and genetic improvement of maize genetic resources. It also provides new insights and presents new perspectives and future research work that has been carried out for further improvement of maize crops. This book is a useful resource for students, researchers, and scientists.

The book includes nine chapters.

Chapter 1, “Silicon, Potassium and Nitrogen Accumulation and Biomass in Corn under Hydroponic Conditions”, evaluates the effect of the interaction of silicon, potassium, and nitrogen on the foliar area, the accumulation of these elements in the aerial part and the dry biomass in corn plants. Chapter 2, “Improved Technological Processes on the Nutritional Quality of Maize,” provides insights into improved techniques such as crossbreeding, genetic cloning, and functional genomics for improving maize nutritional quality. Chapter 3, “Maize (*Zea mays*) Response to Abiotic Stress”, provides useful information on abiotic stress effects on maize crop. Chapter 4, “Climate-Smart Maize Breeding: The Potential of Arbuscular Mycorrhizal Symbiosis in Improving Yield, Biotic and Abiotic Stress Resistance, and Carbon and Nitrogen Sink Efficiency,” discusses important information on climate smart maize breeding. Chapter 5, “Aflatoxins and Fumonisins Contamination of Maize in Bangladesh: An Emerging Threat for Safe Food and Food Security”, provides results that could serve as a benchmark for monitoring mycotoxin contamination in maize. Chapter 6, “Critical Dry Spell Prediction in Rain-Fed Maize Crop Using Artificial Neural Network in Nigeria,” shows how to predict critical dry spells in rain-fed maize crops using artificial neural networks. Chapter 7, “Improving Maize Shelling Operation Using Motorized Mobile Shellers: *A Step towards Reducing Postharvest Losses in Low Developing Countries*” concludes that in addition to other sheller performance attributes, motorized mobile maize shellers can solve transportation challenges associated with motorized immobile maize shellers. Chapter 8, “Advances and Trends in the Physicochemical Properties of Corn Starch Blends,” describes advances and trends in the physicochemical properties of corn starch blends *Zea mays* L. Finally, Chapter 9, “The Impact of Climate Change on Changing Pattern of Maize Diseases in Indian Subcontinent: A Review”, assesses the potential effects of climate change on maize pathogens and consequently on plant health.

The editor would like to thank Publishing Process Managers Iva Ribic and Milica Abeer at IntechOpen for their wholehearted cooperation in the publication of this book.

Mohamed Ahmed El-Esawi, Ph.D.
Botany Department,
Faculty of Science,
Tanta University,
Egypt

Sainsbury Laboratory,
University of Cambridge,
Cambridge, United Kingdom

Chapter 1

Silicon, Potassium and Nitrogen Accumulation and Biomass in Corn under Hydroponic Conditions

Leónides Castellanos González, Renato de Mello Prado and Cid Naudi Silva Campos

Abstract

The aim of the research was to evaluate the effect of the interaction of silicon, potassium, and nitrogen on the foliar area, the accumulation of these elements in the aerial part and the dry biomass in corn plants. The research was developed under hydroponic conditions in Jaboticabal Sao Pablo, Brasil using the 30A77HX hybrid. Two silicon concentrations were evaluated (0 and 2 mmol L^{-1}); two concentrations of potassium (1 and 12 mmol L^{-1}) and four nitrogen concentrations: (1, 10, 15, and 20 mmol L^{-1}). A completely randomized design was used, with factorial arrangement 2 × 2 × 4 and three replications. The foliar area, the dry biomass and, nitrogen, potassium, and silicon content were determined. The application of silicon at a high concentration of K causes an increase in the accumulation of K, which is reflected in an increment of the total dry biomass in the plants of corn, while excess and a deficit of N diminish the accumulation of Si in the aerial part of the plant, which is more evident at a low concentration of K in the nutritious solution, affecting the accumulation of the total dry biomass.

Keywords: benefic element, dry biomass, nutritious, *Zea mays*

1. Introduction

Nitrogen (N) and potassium (K) are essential elements for plants life being part of multiple structural compounds and participating in many vital processes [1]. On the other hand, silicon (Si) is not considered an essential element for plants, however, its absorption can produce beneficial effects in some crops, such as resistance to pests and diseases [2], attenuating abiotic stress [3, 4] and hydric stress [5]. Among the main accumulative crops of Si, are gramineous as sugar cane (*Saccharum* spp.), rice (*Oriza sativa* L.), and corn (*Z. mays* L.) [6].

Some research report about the effect of the combinations of silicon and nitrogen on plants [7–10] and silicon and potassium on the development of plants and yields [11], mainly in accumulative crops of silicon. However, there are not enough results on the effects produced by the interaction of silicon, nitrogen, and potassium on the accumulation of these elements inside the vegetables and their implications on the growth of the plants.

High nitrogen concentration and low concentrations of potassium increase the susceptibility of the corn plants to the noxious agents because it diminishes the absorption of Si. This fact is important because it is well-known that silicon can induce higher resistance to the plants in front of the pests [2].

In an accumulative crop of Si like rice, Andreotti et al. [7], informed that the supply of Si had a small influence on the production of dry matter, although it increased the number of panicles per plant at the highest concentrations of Si, and on the other hand, the concentration of Si decreased with the increment of the dose of fertilization with urea [11].

However, some researchers have stated that there is a lack of information on the use of silicon in general and in particular in corn crops that justifies the need to carry out further research on this subject [12]. Castellanos et al. [13] demonstrated the positive effect of the application of Si on the damages of *Spodoptera frugiperda* Smith in corn to an intermediate dose of N, while Matías and García-Montalvo [14] pointed out that the positive role of the silicon in the resistance to the foliar insects in *Zea diploperennis* L., however in *Z. mayz* this is not yet well explained.

González et al. [15] verified the increment of the green forage in one variety of corn and not in others when the application of an intermediate dose of Si was made, while any varieties had a response at the highest dose.

In front of this situation, two hypotheses arise, one that the unbalanced management of N and K, or the excessive use of N associated with the insufficient application of potassium can diminish the silicon absorption and the production of dry biomass in the plants, and other, that the use of silicon can improve the response of the plant in dry biomass production in function of the application of N and K.

To confirm one of those hypotheses the aim of the present investigation was to evaluate the effect of the interaction of silicon, potassium, and nitrogen on the foliar area and the accumulation of these elements and the dry biomass in corn plants.

2. Material and methods

The investigation was performed in a greenhouse located at the department of soils and fertilizers, FCAV/UNESP Jaboticabal Campus, SP, with geographic coordinates of 21° 15′ 22" South, 48° 18′ 58" West and an elevation of 600 m between March and June of 2014, in a hydroponic floating system.

Two silicon concentrations were evaluated (0 and 2 mmol L^{-1}) using as source silicate of calcium; two concentrations of potassium (1 and 12 mmol L^{-1}) corresponding to 16 and 200% of the solution of K proposed by Hoagland and Arnon [16], using as source monobasic potassium phosphate, and four nitrogen concentrations: 1, 10, 15, and 20 mmol L^{-1} corresponding at 10, 100, 150, and 200% of the solution of Hoagland and Arnon [16], respectively. The added nitrogen corresponded the 25% to ammoniac form (from ammonium chloride) and the 75% to nitric form (from calcium nitrate).

The rest of the macronutrients and micronutrients were incorporated into de nutrient solutions as were proposed by Hoagland and Arnon [16], balancing the concentrations of calcium and phosphorous. The nutrient solution was maintained under continuous oxygenation by means of an air compression system.

Treatments were arranged in one 2 × 2 × 4 factorial scheme with three repetitions. Each experimental unit consisted of a polypropylene pot with a lid, measuring 48 cm long, 11 cm wide at the lower base, 16 cm wide at the upper base, and 17 cm tall, containing 8 L of nutrient solution and six corn plants (Hybrid 30A77HX). The plants were developed in a greenhouse. Initially, the sowing of the corn was carried out in vermiculite on isospory trays, irrigated for 15 days, time in that plants reached five leaves.

Water used in the hydroponic system was distilled and deionized, where solution levels were completed daily in each pot with stock solutions corresponding to each treatment. Values of pH were adjusted to between 6.0 ± 0.2 using solutions of HCl 1.0 mol L^{-1} or NaOH 1.0 mol L^{-1}.

At 45 days after transplanted, the foliar area of the plants was evaluated. For that, all the leaves of the six plants of each pot were collected, being used an integrative apparatus of scanning the foliar area (LI-COR®modelo LI-3000C).

Later on, the dry biomass was determined from the collection of the roots and the aerial part of each pot. For this, the picked-up material was placed in paper bags and dried off in an oven with forced air circulation (65°C) until they reached a constant weight to determine the dry biomass content by pot (aerial part, roots, and the total).

The dry material was ground for chemical analysis of N and K content according to the methodology described by Battaglia et al. [8] and silicon according to Kraska and Breitenbech [17]. Using data of concentration of N, K, and Si in the dry biomass from the aerial part, from the root, and from the total, for each pot, the accumulation of each element per pot was calculated and expressed in mg per plant.

The data of the active foliar area, dry biomass, nitrogen, potassium, and silicon accumulation in the aerial part of the plants were submitted to variance analysis. The media was compared by means of the Tukey test ($P < 0.05$). The statistical package SPSS version 21 was used [18].

3. Results

The application of silicon in the nutritious solution increased the biomass of the plants of corn and the accumulation of N, Si, and K in this, however, the foliar area did not increase, while a high-dose of K caused an increment of all evaluated variables. The foliar area, the biomass of the plants, and the accumulation of the three elements in the plants were influenced in some way by nitrogen dose (**Table 1**). Foliar and total dry matter and the accumulation of K in the plants were influenced by the interaction of the treatments of Si and K, but not the other variables, while

Source of de variation	Accumulation in the aerial part of			Foliar area	Biomass		
	N	Si	K		Roots	Aerial	Total
	Signification of F values						
Si	**	**	**	ns	**	**	**
K	*	**	**	**	**	**	**
N	**	**	**	**	**	**	**
Si × K	ns	ns	**	ns	ns	**	**
Si × N	ns	ns	**	ns	**	ns	ns
K × N	ns	**	**	ns	ns	**	**
CV (%)	14.12	11.51	15.6	14.42	12.51	9.50	8.85

**, **, and ns: Significant ($P < 0.05$), ($P < 0.01$), and nonsignificant by the F-test, respectively.*

Table 1.
Effect of nitrogen (N) and potassium (K) concentrations in the presence or not of silicon (Si), on the nitrogen, potassium, and silicon accumulation in the aerial part of corn plants, the foliar area to and biomass, under hydroponic conditions.

dry biomass of the root and accumulation of K were influenced by the interaction Si × N. The interaction of N and K had influenced on the accumulation of Si and K in the plants and on the increment of the dry biomass of the aerial part of the plants and the total.

The nitrogen accumulation was increased at the highest concentrations of this nutrient with a relationship at the lowest dose while this did not happen at the lowest dose (1 mmol L^{-1}). The silicon accumulation was higher at the concentration at 10 mmol L^{-1} of N, with a statistical difference with the treatments at 15 and 20 mmol L^{-1} of N in the nutrient solution being lower at 1 mmol L^{-1}. A similar situation was observed in relation to the influence of N doses for the foliar area (**Table 2**).

The higher values in accumulation of K were observed at the concentrations of 15 and 20 mmol L^{-1} of N, and the lower at 1 mmol L^{-1}. A similar situation to that it was observed for the total dry biomass, however, the roots dry biomass showed lower values at 1 mmol L^{-1} of N in relation to the highest concentrations. The interaction of the application of Si with the highest concentration of K promoted an increase in the accumulation of K and the foliar and total biomass. However, there was no difference for the foliar biomass between the treatment of Si at 2 mmol L^{-1} combined with the treatment with K at 12 mmol L^{-1} compared with K at 12 mmol L-^{1} without the application of Si (**Table 3**).

The accumulation of K was increased in the treatments that received 2 mmol L^{-1} of Si and 10 and 15 mmol L^{-1} of N, without statistic difference with the treatment that received N at 10 mmol L^{-1} without the application of Si. The dry biomass of the root was increased in the interaction of Si at 2 mmol L-^{1} and N at 10 mmol L-^{1} in relation to the treatment that received 1 mmol L^{-1} of N in the nutrient solution

Nitrogen	**Accumulation in the aerial part of**			**FA**		**Biomass**	
mmolL^{-1}	**N**	**Si**	**K**		**Roots**	**Aerial**	**Total**
		mg kg^{-1}				**g per pot**	
1	70.50b	36.79c	394.14c	0.67c	19.51b	36.50c	56.01c
10	269.32a	71.55a	731.96a	1.20a	29.95a	69.11a	99.06a
15	239.65a	59.62b	665.38ab	1.12b	24.39ab	60.91b	85.30ab
20	252.51a	55.85b	612.37b	1.06b	22.93ab	56.90b	79.83b

Different letters in the columns indicate significant differences according to the Tukey test for $P < 0.05$.

Table 2.
Effect of the nitrogen (N) concentrations on the N, K, and Si accumulation in the aerial part of corn plants, the foliar area (FA), and biomass under hydroponic conditions.

Concentrations		**Accumulation of K**	**Dry biomass**	
Si	**K**	**mg kg^{-1}**	**g per pot**	
mmol L^{-1}			**Aerial**	**Total**
0	1	47.98c	69.63b	246.01c
0	12	57.96b	80.29ab	822.52b
2	1	54.47b	77.02ab	265.48c
2	12	62.22a	88.57a	1069.82a

Different letters in the columns indicate significant differences according to the Tukey test for $P < 0.05$.

Table 3.
Effect of the interaction of nitrogen (N) and potassium (K) concentrations on the N, K and Si accumulation in the aerial part of corn plants, the foliar area and biomass under hydroponic conditions.

without the application of Si, but not in the relation of the rest of the treatments. The role of silicon in the absorption of the K was verified, and at the same time that the beneficial effect of silicon was not evidenced in front of low and high concentrations of N (**Table 4**).

The accumulation of Si and the dry biomass in the air part of the plant manifested an increment at the higher concentration of K (12 mmol L^{-1}) combined with 1, 10, and 15 mmol L^{-1} of N, while the accumulation of K and the biomass of the air part showed increments at 1 mmol L^{-1} of K and 10 mmol L^{-1} of N and at 12 mmol L^{-1} of K and at 10, 15 and 20 mmol L^{-1} of N (**Table 5**).

The influence of the concentration of K in the nutritious solution on the absorption of Si was verified. The application of low and high N concentrations caused the less accumulation of Si in the plant. In the same way, the results demonstrated that the treatments that stood out for a bigger absorption of Si also stood out for higher total biomass, while those treatments that had shown higher accumulation levels of potassium also manifested higher levels of dry biomass in the air part of the plant.

Si	N	Accumulation of K	Roots biomass
Mmol L^{-1}		mg kg^{-1}	g per pot
0	1	318.69f	18.73b
0	10	735.97ab	23.45ab
0	15	513.11de	22.40ab
0	20	569.32cde	23.38ab
2	1	469.58ef	23.30ab
2	10	727.95abc	28.45a
2	15	817.67a	26.38ab
2	20	655.42bcd	22.48ab

Different letters in the columns indicate significant differences according to the Tukey test for P < 0.05.

Table 4.
Effect of the interaction of silicon (Si) and nitrogen (N) concentrations on K accumulation and root biomass under hydroponic conditions.

Concentrations		Accumulation		Biomass	
K	N	Si	K	Aerial	Total
Mmol L^{-1}		mg kg^{-1}		g per pot	
1	1	40.54ef	59.65 cd	44.10c	245.21d
1	10	59.18bcd	83.23ab	65.63ab	246.96d
1	15	51.40 cd	72.73bcd	46.73bc	264.80d
1	20	50.36de	70.83bcd	44.43c	285.05d
12	1	32.48f	52.40d	29.48c	522.26c
12	10	69.05a	96.90a	77.76a	1199.11a
12	15	67.00ab	94.55a	72.52a	1083.79ab
12	20	61.85bc	83.91ab	67.27a	699.53b

Different letters in the columns indicate significant differences according to the Tukey test for P < 0.05.

Table 5.
Effect of the interaction of potassium (K) and nitrogen (N) concentrations on the Si and K accumulation in the aerial part of corn plants and biomass under hydroponic conditions.

4. Discussion

The present results agree with those of Lima et al. [19] who observed increases in dry biomass of leaves, stems, and roots in maize seedlings with the application of 1 mmol L^{-1} of Si via nutrient solution.

Many authors have pointed out the increase of silicon accumulation in different crops as a response to the application of this beneficial element in corn [12, 20] and in rice under hydroponic conditions [11].

The increase of K accumulation in the aerial part of the plant with the application of high concentrations of this nutrient agrees with the results of Andreotti et al. [7] who obtained an increment of the concentration of K with an increase of the doses of K.

Increments in N accumulation as a function of N doses were also obtained in the aerial part of the maize plants (leaves, stems, cob, straw, and grains) by Gava et al. [21]. There are other results that also show that increases in N doses caused higher development and yield of corn plants [21, 22].

The total dry biomass in the aerial part of the plant was increased with the presence of Si, which agrees with the results of Rohanipoor et al. [23], who reported increases in the leaf area of the corn crop under different doses of Si. Also, González et al. [15] observed increases in plant height and green forage of the Morocho Blanco corn variety under hydroponic conditions when they applied an intermediate dose of Si, in relation to nonapplication. They attributed this to a synergism between Si and K.

No results were found in the literature on the influence of the interaction of the presence of Si with a high concentration of potassium on the gain of the dry biomass, however, Miaoo et al. [24] reported that in the soybean plants there was a positive effect of silicon on the increase of root length and its density subjected at low concentration of K (1 mmol L^{-1}). They attributed this to the action of Si on the affectation of the peroxidase enzyme, but this effect was not demonstrated for the case of maize in an experiment in which the plants were subjected to a low concentration of K [10].

The increase in the accumulation of dry matter observed with the application of Si can be associated with its protective effects of the photosynthetic apparatus of the plants, in the improvement of the efficiency of water use and the balance of mineral nutrients as indicated by Mateos-Naranjo et al. [25]. Other researchers have attributed this increment of biomass to the beneficial effects of Si against the oxidative damage of plant membranes and the increase of the cell wall extension capacity [9].

Other investigations have shown that Si increases stomatal conductance by promoting better water use efficiency [26], inducing increased transpiration, which may lead to increased absorption of K, an element that participates actively in the closure and opening of the stomata [25].

The role of Si in the uptake of K was verified in the present research, but this effect does not occur at low and high concentrations of N. The increase of K at 10 mmol L^{-1} of N without application of Si can be explained by the small amounts of Si which remains in the water as indicated by Raya and Aguirre [6], despite being deionized in the experiment.

Parveen and Ashraf [27] also observed increases in dry biomass of the roots in corn plants with the application of 2 mmol L^{-1} of Si in relation to the nonapplication of this element combined with N at 10 mmol L^{-1} under hydroponic conditions.

The results obtained have relation to those of Mauad et al. [11] in rice, who observed a higher concentration of Si in the plant at the lowest dose of N combined with a normal dose of K, independently of the presence or absence of silicon, which show the role of the beneficial element in the presence of plant stress by N.

Castellanos et al. [13] also stated that the interaction of N and K influenced the accumulated silicon in corn with an optimum at 11.4 mmol L^{-1} of N. Similar results were obtained by Mauad et al. [11] who observed a decrease in the deposition of silica in leaves of rice plants at high doses of N.

Increments of dry biomass and productivity of corn as increase the concentration of N have also been obtained by other authors as Queiroz et al. [1].

The nitrogen effects on dry biomass gain was confirmed since this element is a constituent of all molecules, proteins, enzymes, coenzymes, nucleic acids, and cytochromes, as well as its important function as a member of the molecule of chlorophyll, as have been pointed out by Pina et al. [28].

5. Conclusions

The absorption of silicon in corn plants is influenced by the interaction of nitrogen and potassium. Application of Si combined with a high concentration of K causes an increase in K accumulation which is reflected in higher total dry biomass in corn plants. An excess and a deficit of N decrease the accumulation of Si in the aerial part of the plant, which is more evident at a low concentration of K in the nutrient solution, diminishing the accumulation of total dry biomass.

Acknowledgements

To CAPES (Coordenação de Aperfeiçoamento de Pessoal de Nível Superior - Brasil) by the fellowship as visiting professor from abroad granted to the first author.

Author details

Leónides Castellanos González[1]*, Renato de Mello Prado[2]
and Cid Naudi Silva Campos[2]

1 Facultad de Ciencias Agrarias, Universidad de Pamplona,
Pamplona, Norte de Santander, Colombia

2 Faculdade de Ciências Agrárias e Veterinárias, Universidade Estadual Paulista
"Júlio de Mesquita Filho", (UNESP), São Paulo, Brasil

*Address all correspondence to: lclcastell@gmail.com;
leonides.castellanos@unipamplona.edu.co

References

[1] Queiroz AM, Souza CHE, Machado VJ, Quintão RML, Korndorfer GH, Silva AA. Avaliação de diferentes fontes e doses de nitrogênio na adubação da cultura do milho (*Zea mays* L.). Revista Brasileira de Milho e Sorgo. 2011;**10**(3):257-266

[2] Castellanos L, Campos CN, Mello R. Silicon in the crop resistance to agricultural pest. Cultivos Tropicales. 2015a;**36 No especial**:18-26

[3] Silva CN, Renato MR, Caione G, Lima AJ, Checchio FL. Silicon and excess ammonium and nitrate in cucumber plants. African Journal of Agricultural Research. 2016;**11**(4):276-283. DOI: 10.5897/AJAR2015.10221

[4] Mahdieh M, Habibollahi N, Amirjani MR, Abnosi MH, Ghorbanpour M. Exogenous silicon nutrition ameliorates salt-induced stress by improving growth and effciency of PSII in *Oryza sativa* L. cultivars. Journal of. Soil Science and Plant Nutrition. 2015;**15**(4):1050-1060

[5] Camargo MS, Bezerra BKL, Vitti AC, Silva MA, Oliveira AL. Silicon fertilization reduces the deleterious effects of water deficit in sugarcane. Journal of Soil Science and Plant Nutrition. 2017;**17**(1):99-111

[6] Raya Pérez, Juan Carlos, Aguirre Mancilla, César L. El Papel del Silicio en los Organismos y Ecosistemas. Conciencia Tecnológica, núm. 43, enero-junio, 2012, pp. 42-46. Instituto Tecnológico de Aguascalientes, Aguascalientes, México.

[7] Andreotti M, Souza ECA, Crusciol CAC, Rodrigues JD, Büll LT. Produção de matéria seca e absorção de nutrientes pelo milho em razão da saturação por bases e da adubação potássica. Pesquisa Agropecuária Brasileira. 2000;**35**(12):2437-2446

[8] Bataglia OC, Furlani AMC, Teixeira JPF, Furlani PR, Gallo JR. Métodos de análise química de plantas. Campinas: Instituto Agronômico de Campinas; 1983. 48p. (Boletim Técnico, 78)

[9] Jiao-Jing L, Shao-Hang L, Pei-Lei X, Xiu-Juan W, Ji-Gang B. Effects of exogenous silicon on the activities of antioxidant enzymes and lipid peroxidation in chilling-stressed cucumber leaves. Agricultural Sciences in China. 2009;**8**(9):1075-1086

[10] Tewari RK, Kumar P, Tewari N, Srivastava S, Sharma PN. Macronutrient deficiencies and differential antioxidant responses – influence on the activity and expression of superoxide dismutase in maize. Plant Science. 2004;**66**(3):687-694

[11] Mauad M, Costa CA, Grassi Filho CH, Machado SR. Deposição de sílica e teor de nitrogênio e silício em arroz. Semina: Ciências Agrárias. 2013;**34**(4):1653-1662

[12] Ávila FW, Baliza DP, Valdemar F, Lopes JA, Ramos SJ. Interação entre silício e nitrogênio em arroz cultivado sob solução nutritiva 1. Revista Ciência Agronômica. 2010;**41**(2):184-190

[13] Castellanos L, Mello R, Silva GB, Campos CN, Fernández O, Silva R, et al. Daños por *Spodoptera frugiperda* Smith en maíz en función de nitrógeno, potasio y silicio. Revista de Protección Vegetal. 2015b;**30**(3):176-178

[14] Matías G, García-Montalvo IA. Mecanismos de resistencia a patógenos e insectos herbívoros en teosinte y maíz. Journal of Negative & No Positive Results. 2016;**1**(5):190-198

[15] González EM, Ceballos J, Benavides O. Producción de forraje verde hidropónico de maíz *Zea mays.* L.

en invernadero con diferentes niveles de silicio. Revista de Ciencias Agrícolas. 2015;**32**(1):75-83

[16] Hoagland DR, Arnon DI. The Water Culture Method for Growing Plant Without Soil. Berkeley: California Agricultural Experimental Station; 1950. p. 347

[17] Kraska JE, Breitenbeck GA. Simple, robust method for quantifying silicon in plant tissue. Communications in Soil Science and Plant Analysis. 2010;**41**(17):2075-2085

[18] IBM. IBM SPSS Statistics for Windows, Version 21.0. IBM Corp., New York, EE.UU. 2012. ISBN-13: 978-0205985517

[19] Lima M, de Castro VF, Batista J. Aplicação de silício em milho e feijão-de-corda sob estresse salino. Revista Ciência Agronômica. 2011;**42**(2): 398-403.

[20] Barbosa FL, Coelho EM, Mendonçay NCM, Benetoli TR. Adubação foliar com silício na cultura do milho. Revista Ceres, Viçosa. 2011;**58**(2):262-267.

[21] Gava GJC, Oliveira MA, Jerônimo EM, Cruz JCS, Trivelin PCO. Produção de fitomassa e acúmulo de nitrogênio em milho cultivado com diferentes doses de **15**N-uréia. Semina: Ciências Agrárias. 2010;**31**(4):851-862

[22] Victória EL, Fernandes HC, Lacerda EG, Rosado TL. Acúmulo de nutrientes e matéria seca pelo milho em função do manejo do solo e da adubação nitrogenada. Engenharia na Agricultura. 2012;**20**(2):104-111

[23] Rohanipoor A, Norouz IM, Abdolamir Moezzi A, Hassibi P. Effect of Silicon on Some Physiological Properties of Maize (*Zea mays*) under Salt Stress. Journal of Environmental Biology. 2013;7(20):71-79

[24] Miao H, Xing-Guo H, Wen-Hao Z. The ameliorative effect of silicon on soybean seedlings grown in potassium-deficient medium. Annals of Botany. 2010;**105**:967-973

[25] Mateos-Naranjo E, Andrades-Moreno L, Davy AJ. Silicon alleviates deleterious effects of high salinity on the halophytic grass *Spartina densiflora*. Plant Physiology and Biochemistry. 2013;**63**(1):115-121

[26] Farshidi M, Abdolzadeh A, Sadeghipour HR. Silicon nutrition alleviates physiological disorders imposed by salinity in hydroponically grown canola (*Brassica napus* L.) plants. Acta Physiologiae Plantarum. 2012;**34**(5):1779-1788

[27] Parveen N, Ashraf M. Role of silicon in mitigating the adverse effects of salt stress on growth and photosynthetic attributes of two maize (*Zea mays* L.) cultivars grown hydroponically. Pakistan Journal of Botany. 2010;**42**(3):1675-1684

[28] Pina NCA, Herrera OF, Mello R. Manejo de suelos para una agricultura sostenible. 1ra. ed. Vol. 1. Jaboticabal: FCAV/UNESP; 2013. 511 p

Chapter 2

Improved Technological Processes on the Nutritional Quality of Maize

Cebisa Noxolo Nesamvuni, Khavhatondwi Rinah Netshiheni and Oluwaseun Funmi Akinmoladun

Abstract

As global food security and staple food, maize has become one of the most widely used cereals for fundamental research. Several important discoveries are reported, some of which are technological processes being used to improve maize crops' dietetic, phenotypic, genotypic, and organoleptic properties. This chapter provides insight into improved technological techniques such as crossbreeding, genetic cloning, and functional genomics and how they improve the nutritional quality of maize crops. The use of these technological processes could be one of the sustainable strategies in meeting the dietary needs and livelihood of Africa, Mexico, and Latin America's growing populace.

Keywords: breeding, genomics, functional genomics, improved technological processes, maize nutritional quality

1. Introduction

Maize (*Zea mays ssp.*) is one of the widely-spread staple cereals globally since its introduction to the New World by Christopher Columbus in the fifteenth century [1, 2]. Maize originated from central Mexico 7000–9000 years ago as a wild grass known as teosinte [3]. Today, maize is a cereal that serves as a significant food source in animal and human nutrition, playing an essential role in feeding the world. It is the most researched cereal due to its significant strategic role in social and economic development, mainly in Asia and Africa [4], impacting economic growth activities, including employment. Based on FAOSTAT [5] report, maize and its products contributed 6.5%, 30%, and 38% of food supply to Asia, the Americas, and Africa. Africa's farmland used for maize cultivation is 24% [5]. Although maize production has made a critical contribution to food security and poverty in many African countries, there have been persistent challenges, causing a low maize yield and poor crop nutritional quality.

Nevertheless, Otekunrin et al. [6] assert that maize production can still play an essential role in achieving poverty alleviation and zero hunger. They illuminated the importance of channeling the support from the agricultural ministry on educated maize farmers from an empirical study conducted in Ghana. They argue that knowledgeable farmers can effectively use new technologies. Technical efficiency, which is the ability to use available resources for maximum output, influences the choices for the strategies used for productivity improvement.

Maize evolved enormously, alienating itself from some key traits of teosinte. For example, teosinte has abundant branches and tillers, increased number of ears per plant, reduced number of kernels per ear (5–12 per ear for teosinte and several hundred for maize), and small kernels with a hardened fruit case (reviewed in [7]). About 40 years ago, Beadle [8] observed that after he planted a teosinte–maize F_2 population consisting of 50,000 individuals, the frequency of parental types was ~1 in 500, then estimated that there were four or five major loci involved in maize domestication. Later, Doebley and Stec [9] mapped five major quantitative trait loci (QTLs) plus some minor-effect QTLs for key traits in which teosinte and maize differ. This result was consistent with Beadle's estimation and indicated that a small number of loci were responsible for the teosinte–maize morphological difference. Wright [10] investigated 774 genes and estimated that 2–4% of maize genes were selected during maize domestication and subsequent improvement. According to recently released gene annotations of high-quality maize genomes, modern maize contains ~39,000–42,000 protein-coding genes [11–14], indicating that ~800–1700 protein-coding genes (39,000 × 2% = 800; 42,000 × 4% = 1700) underwent selection during the process of domestication (**Figure 1**).

Recently, two researchers [16, 17] used chromatin interaction analysis by paired-end tag sequencing (ChIA-PET) technology to map genome-wide chromatin interactions. They revealed their connections to gene-expression regulation (**Figure 1**), including the chromatin interactions in which TB1, UB3, ZmCCT9, Vgt1 were involved. Likely, population-scale identification of chromatin interactions would allow the detection of many more important regulatory elements, providing several useful selection targets for improving future maize, which is essential in addressing poverty alleviation and zero hunger (sustainable development goals, 1 and 2, set by the United Nations in 2015 to achieve global food security by 2030) [18]. The current low in the maize crop yield calls for more comprehensive and more consistent crop production strategies. A newly introduced program named clustered regularly interspaced short

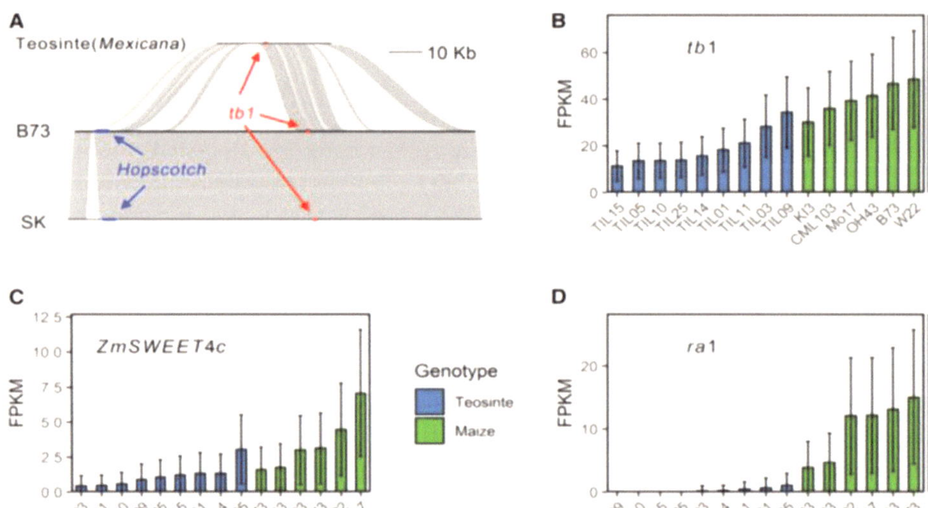

Figure 1.
Genomic sequence variants in the tb1 regions of teosinte and tropical and temperate maize lines [1]. (A) The red rectangles indicate the position of tb1, and the blue rectangles indicate the position of the hopscotch TE. This TE is the functional variant of tb1 and is absent in teosinte [14]. (B–D) the increased expression levels of representative selected genes (tb1 in B, ZmSWEET4c in C, ra1 in D) in modern elite maize lines compared with teosinte, the ancestor of maize. The expression profile was obtained by analyzing RNA-seq data generated by Lemmon et al. [15].

palindromic repeats-associated protein (CRISPR-Cas) technology is widely used for plant genome editing. It can be hoped that the CRISPR-Cas system will accelerate the breeding of improved crop cultivars compared with conventional breeding and help to address the zero-hunger goal.

Although maize protein is high in the ratio of leucine to isoleucine, it lacks tryptophan and lysine, making it poor nutritionally. Also, threonine is found in a reduced quantity in common maize [19]. However, Mertz et al. [20] discovered an improved nutritional quality of maize mutant called *opaque* 2 (*O2*). This *O2* maize has 95% casein and 43% higher protein quality than common maize. Hence, efforts were made to integrate *O2* as a commercialized variety but hindered by processing and agronomic problems [21]. *O2* was characterized by the dull and chalky kernel, reduced grain yield, and susceptibility to stored grain pests and soft endosperm, making it unacceptable to farmers and consumers. Hence, quality protein maize (QPM) emerged. QPM is a genotype that incorporated modified *opaque 2.* QPM improved the poor keeping quality, deficient agronomic attributes of *opaque 2*, and the truncated nutritive value of normal endosperm; hence, it contained double tryptophan and lysine when juxtaposed with common maize endosperm [22]. Conventional breeding was used to develop QPM that possessed high lysine and complex endosperm characters by International Centre for Maize and Wheat Research (CIMMYT), Mexico, in 1993.

The breeding of QPM was introduced to improve the nutritional composition of protein in maize grain. Maize seeds contain an alcohol-soluble protein called zein [19]. QPM has a protein profile of 90% milk protein compared to the 40% milk biological value protein found in common maize [19]. Zein is more in the endosperm than in the embryo and constitutes 50–70% endosperm protein. Zeins are high in leucine, proline, and glutamine but are lack in tryptophan and lysine. Therefore, zein compositions are altered to enhance maize nutritional quality [19].

Maize products are shaped to make nutritious foods more available by using desirable characteristics and traits. Therefore, new varieties with high yields became the focus of maize breeders [23]. Information on the needs of maize users can be incorporated into the products' characteristics by the breeders. This information will increase the use of maize varieties and, most importantly, improve nutrition [24]. Worldwide, maize varieties vary genetically by hardness, sweetness, and grain's size and color, and this genetic variation results in diversity in nutritional properties of the whole maize grain. Maize endosperm is made up of 82% composition of maize kernel, which is mainly starch. Hence, the endosperm' protein profile of the maize kernel was improved by the QPM. The breeding of this QPM has been achieved through alteration in the recessive mutant allele of the *O2* gene, a specific set of amino acid modifier genes and pair of endosperm hardness modifier genes. The kernels can be flint, pop, waxy, floury, or dent and provide micronutrients and macronutrients. There is a high level of antioxidants and minerals in the aleurone; minerals and fiber in the pericarps; antioxidants, protein starch, and vitamins in the endosperm; vitamins like vitamin E, fat, and minerals are rich in the germs [25]. The primary compounds in the kernels are cellulose and lignin, while secondary compounds are hemicellulose, β-glucans, and arabinoxylans. In maize, the presence of phytochemicals (anthocyanins, phlobaphenes, carotenoids, phenolic acids, nonpolar and polar lipids) prevents diseases and strengthens health.

2. Technologies on the nutritional quality of maize crop

This technological processing for increasing the nutritional quality of maize as a potential solution for nutritional deficiency can be classified into two main groups, preharvest technology and postharvest technology.

2.1 Preharvest technology

Preharvest technology discussion will include crossbreeding and genetic manipulation, functional genomics and transgenic crop technology, biofortification, functional genomics and transgenic crop technology, and soil improvement.

2.1.1 Crossbreeding and genetic manipulation

Crossbreeding method is used to transfer micronutrients density in unaccustomed sources into genetic with a high-yielding competitive background. For the farmers to accept and adopt the newly developed trait, end-use quality and agronomic attributes must be considered during crossbreeding [26]. An increase in protein content such as methionine, tryptophan, and lysine is evidence for advancement in diet's nutritional content through breeding that fundamentally focuses on nutritional quality.

Physical appearance, cooking and eating quality, milling trait, and nutritive value are those parameters used to determine grain's overall quality [18]. These properties are to a large extent, especially the eating and cooking qualities, influenced by the amylose content of the maize, and thus open to genetic manipulation. Amylose, a polysaccharide that is made up of 20–30% starch, is a significant form of resistant starch. However, the amylose biosynthesis is modulated by the enzyme granule-bound starch synthase 1 (GBSS1) and encoded by the Waxy gene. Gao et al. [27], and Wang et al. [28] disrupted the GBSS1 with clustered interspaced and short palindromic repeats-associated protein (CRISPR-Cas) to produce elect maize that contains low amylose content. Also, CRISPR-Cas to encode isoamylase-type debranching enzyme been edited with isoamylase 1 (ISA1) produced low amylose content [29]. CRISPR-Cas uses the system of knock-out and knock-in to improve the quality of crops.

Phytate, an antinutrient content in maize, also referred to as inositol 1,2,3,4,5,6-hexakisphosphate, forms insoluble complexes with minerals and protein and reduces their absorption when consumed. Zinc finger nucleases (ZFNs) blocked gene coding IPK1 for enzyme inositol-1,3,4,5,6-pentakisphosphate 2-kinase to reduce the phytate concentration in maize [30]. Also, Qi et al. [31] used CRISPR-Cas and RNA interference (RNAi) which targeted the gene ZmMADS47 encoding a MADS-box protein that interacts with *O2* to switch on zein gene promoter so that reduced zein protein content can be reduced. However, the decrease in zein content was 12.5% and 16.8% in the kernel of MADS/Cas9-21 and ZmMADS47 lines, respectively [31].

2.1.2 Functional genomics and transgenic crop technology

Genome editing is the principle on which functional genomics is based. It is based on nuclease-based forms of engineering like transcription activator-like effector nucleases (TALENTS), clustered regularly interspaced short palindromic repeats (CRISPR) with the concerns of creation of mutations, precise incisions, and substitutions in eukaryotic and plant cells [32]. Transgenic crop technology directly inserts genes of interest into the plant genome. These are the only viable alternative for biofortifying crops with micronutrients that naturally do not exist in the crop [26]. Transgenic crop technology can be achieved at a low cost, short time, and without nutrition-based programs and ease the concurrent incorporation of the genetic system to reduce antinutrients, increase micronutrient concentration, and promote bioavailability.

A hybrid of QPM and provitamin A was developed by Zunjare et al. [33], which was speculated to help fight malnutrition among the populace where maize is used as a primary staple food. Based on QPM analysis, the required amount of tryptophan and lysine was achieved when switching conventional maize with a lesser amount of QPM [34]. Also, orange maize improved vitamin A in children's diet in Zambia [35].

Many kinds of cereal contain a high prevalence of phytic acid (PA), a significant zinc absorption inhibitor. Minerals like iron and zinc are bound by PA and prevent mineral absorption in the gastrointestinal tract. Some researchers like Brnić et al. [36] have reduced the PA in maize due to its nutritional consequences. Chemicals such as acetic acid and hydrochloric acid, microwave treatment or heat methods, and recombinant microbial phytate exogenously reduce PA in grains [37]. Transgenic corn expression phytate from *Aspergillus niger* was created to raise mineral availability by limiting PA through microbial phytate enzymes [38]. Eventually, low PA (lpa) phenotype cereal mutants have been developed in wheat, maize, and rice. The primary concern about this transgenic expression is its negative impact on agronomic performance and crop yield.

2.1.3 Biofortification

Biofortification is a primary means of fighting micronutrient deficiency in the world. Biofortification and supplementation are the traditional means of adding minerals and vitamins to food crops. The three essential means of biofortifying crops are biotechnology, conventional plant breeding, and adding an inorganic or mineral compound to fertilizer [37]. Essential micronutrients and β-carotene have been used to biofortified maize in other to maintain healthy living. In the 1970s, researchers developed quality protein maize (QPM) to increase maize's tryptophan and lysine content. These newly developed biofortified crops have a tremendous amount of micronutrients in the edible parts of the crops. Conventional breeding in maize has also been used to upgrade its nutritional content [25].

Nutritionally improved crops, such as orange maize (enhanced with zinc and biofortified with provitamin A carotenoids), quality protein maize (QPM) (biofortified with amino acids), have been developed by plant breeders. The commercialization of maize with provitamin A carotenoid is gaining traction in Western and Southern Africa [38]. Sowa et al. [39] reported 85% retention of carotenoid provitamin A in biofortified flour in the preparation of muffin and porridge. Adoption of biofortified crops is mainly tested by consumer acceptability of rural households where it is used to produce the different menus and prepared in their local ways. Consumers need to accept and use biofortified crops to prepare local foods to make the most of them. However, QPM was preferred by rural mothers in Ethiopia to conventional maize in the preparation of complementary food for children and infants [40]. Li et al. [41] and Muzhingi et al. [42] proved that biofortified maize efficiently converted provitamin A carotenoid to vitamin A. Hence, the daily requirement of vitamin A may be met by consuming biofortified maize.

Although the stability of carotenoids during storage is a significant challenge in provitamin A maize, consumption without dehulling improved the storage stability of carotenoids. Taleon et al. [43] studied different processing methods and storage environments of hybrid maize in Zambia. They proposed that maize biofortified with provitamin A should be sold just before consumption as whole grain and milled as such. Carotenoid degrades over time; therefore, fortified maize variety should be consumed before the white-unfortified maize.

The milling process, mode of cooking (refined versus whole grain flour), and container used for cooking will determine the zinc retention in biofortified maize. Zinc absorbed by Zambia children through zinc-biofortified maize helped to meet their zinc requirement [44]. QPM boosts tryptophan; hence, niacin (vitamin B3) can be partially met [25] because tryptophan is free by changing the leucine—isoleucine ratio for niacin biosynthesis.

2.1.4 Soil improvement

Protein content and yield react positively to nitrogen fertilizer used to supplement the soil in which maize was planted. This is the advantage of adding nitrogen fertilizer to low-nitrogen soil. Fertilizers that are biofortified with micronutrients and applied to the soil are the simplest biofortification method [37]. This practice has been affected by the regular application of micronutrients to the soil, increasing labor and cost, accumulation and mobility of minerals among plants, and variations in soil composition at a specific location. It also increases the micronutrients temporarily as there is a need to apply the fertilizer constantly. These micronutrient fertilizers raise molybdenum, nickel, copper, selenium, iodine, and zinc in different levels in their edible parts [26].

People in low-income countries whose diet is based on cereal grains low in zinc (Zn) are affected by Zn deficiency. Zn deficiency can cause poor immunity, birth complications, impaired mental development, and stunted growth [37]. Zinc concentration in grain crops has been improved by nitrogen management improvement through the availability of Zn in the soil. Nitrogen availability in the soil represents a significant component in the biofortification of zinc in grains and, therefore, enhances residence's nutritional status in developing countries. Another micronutrient in low quantity in grain due to its insufficient amount in the soil is selenium (Se) [45]. The organic form of Se (selenocysteine and selenomethionine) are more significantly bioavailable than inorganic selenium (Se). According to Poblaciones et al. [45], Se-rich fertilizer increases Se's bioavailability in grain and boosts total yield. The author discovered that chickpea could store a high concentration of Se in the grain after applying fertilizer.

2.2 Postharvest technology

Reducing the wastage and losses of food becomes crucial in ensuring adequate nutrition, food security, improvement in rural livelihood, poverty, and food availability among the populace. However, food wastage can result from poor grain storage affected by temperature, relative humidity and grain moisture. Microorganisms, insects, and rodents are maize biological deteriorating agents [25]. *Penicillium, Fusarium* and *Aspergillus* are the general mycotoxigenic fungi of importance in maize. The most prominent mycotoxins in foods, aflatoxin is produced as secondary metabolites by *Aspergillus flavus sp*. The contribution of aflatoxins to crops losses has a negative effect, either directly or indirectly, on general nutrition, health, food security, and the economy at large [46]. The prevalence of aflatoxin can be reduced by both post and pre-harvest intervention. The preharvest intervention could be in the form of developing insect and Aspergillus resistant, heat, and drought-tolerant varieties [47]. According to Suwarno et al. [48], provitamin A carotenoid enriched maize can reduce aflatoxin contamination. Postharvest interventions include suitable moisture at harvest, humidity and temperatures during storage, and suitable containers and space.

2.2.1 Processing and utensils used

When maize is processed into staple foods, it lacks tryptophan, lysine, and methionine [34]. Different cooking and processing methods, degermed and decorticated kernels in maize products cause additional loss of nutrients. Some processing methods and menus where maize is served as raw materials can still enhance the nutritional properties of the final product and overcome nutrient deficiencies [25]. Fermentation and nixtamalization as processing methods increase bioavailability and bioaccessibility of maize and can also cause a decrease in some compounds.

2.2.2 Unrefined grains

These are grains that their germ and bran had not been removed through processing methods. They help to reduce the risk of type 2 diabetes, heart disease. Overall mortality is positively associated with high unrefined grain and nonfiber consumption. According to Willett et al. [49], preferred foods and shelf-life are the social and technological challenges faced when consuming whole grain. Most consumers of these products have adapted the fine texture and color of most refined flour, especially maize. However, there is a need for promotion and expansion in situations where unrefined maize has been consumed to increase its nutritional impact and meet consumer preferences [50]. These situations include roasted kernels, green maize, popcorn, wet-ground foods, and nixtamalization, along with short soaking time. The aleurone, germ, and pericarp of refined maize have been removed together with minerals and vitamin B. If this maize is not enriched after milling, the consumer's nutritional status will be negatively affected. There are little or no changes in the nutrient content of refined maize if it has been biofortified. Gannon et al. [51] compared the bioavailability of provitamin A from refined and whole maize grain with biofortified maize flours as there was no difference between their β-carotene bioefficiency. Furthermore, milling does not affect the bioavailability of zeaxanthin and β-cryptoxanthin in refined and whole-grain biofortified orange maize [52].

2.2.3 Fresh maize

Any variety of maize harvested at the milky stage and prepared by boiling or roasting is referred to as fresh or green maize. Compared to conventional maize, QPM has higher glutelins, reduced zein protein, peak tryptophan, and lysine [25]. Alamu et al. [53] reported variations in the minerals and macronutrients retention of boiled and fresh maize in yellow, white, and high provitamin A carotenoid. The boiling of maize preserved lysine, zinc, and carotenoids at the milk stage.

2.2.4 Nixtamalization

Nixtamalization is when lime, grounded shellfish shell, or ash is added to hot water in which whole grains have been soaked for 8 hours or more. Any process used to modify grain fat components, remove the pericarp, and produce changes in starch and protein content of any grain is called nixtamalization [50]. There is a better characteristic of the kernels of processed maize through nixtamalization as compared to unprocessed ones. Nixtamalization causes physicochemical changes (losses in the pericarp) in maize kernels because of changes in the functionality and chemical composition (decrease in phytic acid) [25]. Nixtamalized kernel is characterized by reduced mycotoxin content, increased product shelf-life and nutritional value, easy mill kernel, and improved aroma and flavor. The significant nutritional changes that occur because of the nixtamalization process in maize and its products include greater bioavailability of iron and niacin, increased resistance starch content and calcium [25].

2.2.5 Fermentation

Latin America and African countries have a lot of maize food products that are acidic and nonalcoholic fermented. Marco et al. [54] stated that probiotics from fermented products promote a healthy microbiome. Lacto-fermentation is a process by which starch and sugar are converted to lactic acid by bacteria. The nutritional bioavailability of niacin and iron from beverages is increased by fermentation [55]. Furthermore, mycotoxins, antinutrients, and natural toxicants are reduced or removed through fermentation processing, thus improving the maize products' safety and nutritional quality [56].

3. How technological processes improve the nutritional quality of maize crops

Based on the report of Mertz et al. [20], opaque 2 *(O2)* lysine content in the endosperm (3.3–4.4 g lysine/100 g of endosperm) is double that of the normal maize (1.3 glysine/100 g endosperm protein). The *O2* maize protein has a biological value of 90% of milk protein, while normal maize protein has 40%. The body utilized 74% of *O2* maize protein intake, while only 37% was used in normal maize protein [19].

In comparing the protein content of QPM with normal maize, the QPM protein contains 38% lesser leucine, 55% higher tryptophan, and 30% higher lysine than normal maize. Bressani [57] reported that 8 g/kg body weight of QPM is needed for nitrogen equilibrium compared to the 24 g/kg body weight of normal maize. The QPM has greater niacin availability due to lower leucine content, utilization of carotene, and higher tryptophan content [19]. QPM maize can be processed without a decrease in its acceptability and quality. Bressani [57] report in Columbia showed that *O2* maize was used as therapeutic food for children suffering from protein deficiency diseases and kwashiorkor and brought about normalcy in them. Also, Anon [58] reported that QPM increased the weight and height of preschool children that used it as a major starchy staple by 20% faster than those that used normal maize. Based von the report of Muzhingi et al. [42], porridge made from biofortified yellow maize provide 40–50% vitamin A of recommended dietary allowance (RDA) for Zimbabwean men. Similarly, North American females consuming porridge from biofortified maize had 3:1 fold of vitamin A equivalence when a fraction of the blood triglycerol-rich lipoprotein was measured compared to the traditional white maize porridge [41].

The high amount of ascorbate, folate, and β-carotene in triple-vitamin fortified maize has been developed through metabolic engineering in the endosperm [59]. The transgenic kernels have a double, 6-fold and 169-fold normal amount of folate, ascorbate, and β-carotene as traditionally bred crops. These crops can offer a nutritionally complete meal. There was a grain yield of about 145.3% in transgenic maize compared to wild maize due to upgraded grain number and size [60]. Total starch content was improved by constitutive expression of invertase in the transgenic kernels, which showed that genes could boost grain quality and yield in crop plants.

4. Conclusion

Malnutrition and hunger alleviation can be achieved through fortified maize and continuous advancement in crop yields. It is not enough to generate micronutrients at a higher level in plants. Their bioavailability, absorption, and utilization in the body are crucial to increase the consumer's micronutrient status after cooking and processing the food in their local ways. Also, the biofortified crops must be accepted by consumers and adopted by a significant number of farmers to increase the nutritional health of the community. Biotechnology through biofortified maize can be used to improve the vitamin A status of the populace.

Author details

Cebisa Noxolo Nesamvuni, Khavhatondwi Rinah Netshiheni*
and Oluwaseun Funmi Akinmoladun
Department of Nutrition, Faculty of Health Sciences, University of Venda, Thohoyandou, Limpopo Province, South Africa

*Address all correspondence to: khavhatondwi.netshiheni@univen.ac.za; kr.nesh@gmail.com

References

[1] Liu H, Luo X, Niu L, Xiao Y, Chen L, Liu J, et al. Distant eQTL and non-coding sequences play critical roles in regulating gene expression and quantitative trait variation in maize. Molecular Plant. 2020;**10**:414-426

[2] Brandolini A, Brandolini A. Maize introduction, evolution and diffusion in Italy. Maydica. 2009;**54**:233-242

[3] Matsuoka Y, Vigouroux Y, Goodman MM, Sanchez J, Buckler E, Doebley J. A single domestication for maize shown by multilocus microsatellite genotyping. Proceedings of the National Academy of Sciences. 2002;**99**(9):6080-6084

[4] Awika JM. Major cereal grains production and use around the world. In: Advances in Cereal Science: Implications to Food Processing and Health Promotion. Washington, DC: American Chemical Society; 2011. pp. 1-13

[5] FAOSTAT. Food Balance Sheets. 2019. Retrieved from: http://www.fao.org/faostat/en/#data/FBS

[6] Otekunrin OA, Otekunrin OA, Momoh S, Ayinde IA. How far has Africa gone in achieving the zero hunger target? Evidence from Nigeria. Global Food Security. 2019;**22**:1-2

[7] Doebley J. The genetics of maize evolution. Annual Review Genetics. 2004;**38**:37-59

[8] Beadle GW. The mystery of maize. Field Museum of Natural History bulletin. 1972;**43**:2-11

[9] Doebley J, Stec A. Inheritance of the morphological differences between maize and teosinte: Comparison of results for two F2 populations. Genetics. 1993;**134**:559-570

[10] Wright SI, Bi IV, Schroeder SG, Yamasaki M, Doebley JF, McMullen MD, et al. The effects of artificial selection on the maize genome. Science. 2005;**308**:1310-1314

[11] Jiao Y, Peluso P, Shi J, Liang T, Stitzer MC, Wang B, et al. Improved maize reference genome with single-molecule technologies. Nature. 2017;**546**:524-527

[12] Springer NM, Anderson SN, Andorf CM, Ahern KR, Bai F, Barad O, et al. The maize W22 genome provides a foundation for functional genomics and transposon biology. Nature Genetics. 2018;**50**:1282-1288

[13] Sun S, Zhou Y, Chen J, Shi J, Zhao H, Zhao H, et al. Extensive intraspecific gene order and gene structural variations between Mo17 and other maize genomes. Nature Genetics. 2018;**50**(9):1289-1295

[14] Yang N, Wu S, Yan J. Structural variation in complex genome: Detection, integration and function. Science China Life Sciences. 2019;**62**:1098-1100

[15] Lemmon ZH, Bukowski R, Sun Q, Doebley JF. The role of cis regulatory evolution in maize domestication. PLOS Genetics. 2014;**10**:e1004745

[16] Li C, Li W, Zhou Z, Chen H, Xie C, Lin Y. A new rice breeding method: CRISPR/Cas9 system editing of the Xa13 promoter to cultivate transgene-free bacterial blight-resistant rice. Plant Biotechnology Journal. 2020;**18**:313

[17] Peng Y, Xiong D, Zhao L, Ouyang W, Wang S, Sun J, et al. Chromatin interaction maps reveal genetic regulation for quantitative traits in maize. Nature Communications. 2019;**10**:1-1

[18] Ahmad S, Tang L, Shahzad R, Mawia AM, Rao GS, Jamil S, et al. CRISPR-based crop improvements: A way forward to achieve zero hunger. Journal of Agricultural and Food Chemistry. 2021;**69**:8307-8323

[19] Agrawal PK, Gupta HS. Enhancement of protein quality of maize using biotechnological options. Animal Nutrition and Feed Technology. 2010;**10**:79-91

[20] Mertz ET, Bates LS, Nelson OE. Mutant genes that changes protein composition and increases lysine content of maize endosperm. Science. 1964;**145**:279-280

[21] Sun Y, Jiao G, Liu Z, Zhang X, Li J, Guo X, et al. Generation of high-amylose rice through CRISPR/Cas9-mediated targeted mutagenesis of starch branching enzymes. Frontier in Plant Science. 2017;**8**:298

[22] Mertz ET. Discovery of high lysine, high Tryptophan cereals. In: Mertz ET, editor. Quality Protein Maize. Ethiopia: American Society of Cereal Chemistry; 1992. pp. 1-8

[23] Babu R, Nair SK, Kumar A, Venkatesh S, Sekhar JC, Singh NN. Two-generation marker-aided backcrossing for rapid conversion of normal maize lines to quality protein maize (QPM). Theoretical and Applied Genetics; 2009;**111**. 888-897

[24] Holmes M, Renk J, Coaldrake P, Kalambur S, Schmitz C, Anderson N. Food-grade maize composition, evaluation, and genetics for masa-based products. Crop Science. 2019;**59**:1392-1405. DOI: 10.2135/cropsci2018.10.0605

[25] Ekpa O, Palacios-Rojas N, Kruseman G, Fogliano V, Linnemann AR. Sub-Saharan African maize-based foods: Technological perspectives to increase the food and nutrition security impacts of maize breeding programmes. Global Food Security. 2018;**17**:48-56

[26] Palacios-Rojas N, McCulley L, Kaeppler M, Titcomb TJ, Gunaratna NS, Lopez-Ridaura S, et al. Mining maize diversity and improving its nutritional aspects within agro-food systems. Comprehensive Reviews in Food Science and Food Safety. 2020;**19**:1809-1834

[27] La Frano MR, de Moura FF, Boy E, Lönnerdal B, Burri BJ. Bioavailability of iron, zinc, and provitamin A carotenoids in biofortified staple crops. Nutrition Reviews. 2014;**72**:289-307

[28] Gao H, Gadlage MJ, Lafitte HR, Lenderts B, Yang M, Schroder M. Superior field performance of waxy corn engineered using CRISPR–Cas9. Nature Biotechnology. 2020;**38**:579-581

[29] Wang W, Wei X, Jiao G, Chen W, Wu Y, Sheng Z, et al. GBSS-BINDING PROTEIN, encoding a CBM48 domain-containing protein, affects rice quality and yield. Journal of Integrated Plant Biology. 2020;**62**:948-966

[30] Ku HK, Ha SH. Improving nutritional and functional quality by genome editing of crops: Status and perspectives. Frontiers in Plant Science. 2020;**11**:1514

[31] Qi W, Zhu T, Tian Z, Li C, Zhang W, Song R. High-efficiency CRISPR/Cas9 multiplex gene editing using the glycine tRNA-processing system-based strategy in maize. Biotechnology. 2016;**16**:58. DOI: 10.1186/s12896- 016-0289-2

[32] Gaj T, Charles A, Gersbach CA, Barbas CF III. ZFN, TALEN, and CRISPR/Cas-based methods for genome engineering. Trends in Biotechnology. 2013;**31**:397-405

[33] Zunjare RU, Hossain F, Muthusamy V, Baveja A, Chauhan HS, Bhat JS, et al. Development of biofortified maize hybrids through

marker-assisted stacking of *β-carotene hydroxylase, lycopene-ε-cyclase* and *opaque2* genes. Frontiers in Plant Science. 2018;**9**:178. DOI: 10.3389/fpls.2018.00178

[34] Nuss ET, Tanumihardjo SA. Quality protein maize for Africa: Closing the protein inadequacy gap in vulnerable populations. Advances in Nutrition. 2011;**2**:217-224

[35] Gannon BM, Kaliwile C, Arscott S, Schmaelzle S, Chileshe J, Kalungwana N. Biofortified orange maize is as efficacious as a vitamin A supplement in Zambian children even in the presence of high liver reserves of vitamin A: A community-based, randomized placebo-controlled trial. American Journal of Clinical Nutrition. 2014;**100**:1541-1550

[36] Brnić M, Wegmüller R, Zeder C, Senti G, Hurrell RF. Influence of phytase, EDTA, and polyphenols on zinc absorption in adults from porridges fortified with zinc sulfate or zinc oxide. Journal of Nutrition. 2014; **144**:1467-1473

[37] Hefferon KL. Nutritionally enhanced food crops; progress and perspectives. International Journal of Molecular Sciences. 2015;**16**:3895-3914

[38] Andersson M, Saltzman A, Singh Virk P, Pfeiffer W. Progress update: Crop development of biofortified staple food crops under HarvestPlus. African Journal of Food, Agriculture, Nutrition and Development. 2017;**17**:11905-11935

[39] Sowa M, Yu J, Palacios-Rojas N, Goltz SR, Howe JA, Davis CR, et al. Retention of carotenoids in biofortified maize flour and *β*-cryptoxanthin enhanced eggs after household cooking. Omega. 2017;**2**:7320-7328. DOI: 10.1021/acsomega.7b01202

[40] Gunaratna N, Bosha T, Belayneh D, Fekadu T, De Groote H. Women's and children's acceptance of biofortified quality protein maize for complementary feeding in rural Ethiopia. Journal of the Science of Food and Agriculture. 2016;**96**:3439-3445

[41] Li SS, Nugroho A, Rocheford T. Vitamin A equivalence of the β-carotene in β-carotene-biofortified maize porridge consumed by women. American Journal of Clinical Nutrition. 2010;**92**:1105-1112

[42] Muzhingi T, Gadaga TH, Siwela AH, Grusak MA, Russell RM, Tang G. Yellow maize with high β-carotene is an effective source of vitamin A in healthy Zimbabwean men. American Journal of Clinical Nutrition. 2011;**94**:510-519

[43] Taleon V, Mugode L, Cabrera-Soto L, Palacios-Rojas N. Carotenoid retention of biofortified maize in different postharvest storage and packaging methods. Food Chemistry. 2017;**232**:60-66

[44] Chomba E, Westcott CM, Westcott JE, Mpabalwani EM, Krebs NF, Patinkin ZW, et al. Zinc absorption from biofortified maize meets the requirements of young rural Zambian children. Journal of Nutrition. 2015;**145**:514-519

[45] Poblaciones MJ, Rodrigo S, Santamaria O, Chen Y, McGrath SP. Selenium accumulation and speciation in biofortified chickpea (*Cicer arietinum* L.) under Mediterranean conditions. Journal of Science Food Agriculture. 2014;**94**:1101-1106

[46] Hoffmann V, Jones K, Leroy JL. The impact of reducing dietary aflatoxin exposure on child linear growth: A cluster randomized controlled trial in Kenya. British Medical Journal Global Health. 2018;**3**:e000983. DOI: 10.1136/bmjgh-2018- 000983

[47] Mahuku G, Lockhart BE, Wanjala B, Jones MW, Kimunye JN, Stewart LR. Maise Lethal Necrosis (MLN), an

emerging threat to maize-based food security in sub-Saharan Africa. Phytopathology. 2015;**105**:956-965

[48] Suwarno W, Hannok P, Palacios-Rojas N, Windham G, Crossa J, Pixley KV. Provitamin A carotenoids in grain reduce aflatoxin contamination of maize while combating vitamin A deficiency. Frontiers in Plant Science. 2019;**10**:30

[49] Willett W, Rockström J, Loken B, Springmann M, Lang T, Vermeulen S, et al. Food in the Anthropocene: The EAT–Lancet Commission on healthy diets from sustainable food systems. Lancet. 2019;**393**:447-492

[50] Ekpa O, Palacios-Rojas N, Kruseman G, Fogliano V, Linnemann A. Sub-Saharan African maize-based food: Processing practices challenges and opportunities. Food Reviews International. 2019;**35**:696-639

[51] Gannon BM, Pixley KV, Tanumihardjo SA. Maize milling method affects growth and zinc status but not provitamin A carotenoid bioefficacy in male Mongolian gerbils. Journal of Nutrition. 2017;**147**:337-345

[52] Titcomb TJ, Sheftel J, Sowa M, Gannon BM, Davis CR, Palacios-Rojas, et al. *β*-Cryptoxanthin and zeaxanthin are highly bioavailable from whole-grain and refined biofortified orange maize in humans with optimal vitamin A status: A randomized crossover placebo-controlled trial. American Journal of Clinical Nutrition. 2018;**108**:793-802

[53] Alamu O, Menkir A, Maziya-Dixon B, Olaofe O. Effects of husk and harvest time on carotenoid content and acceptability of roasted fresh cobs of orange maize hybrids. Food Science and Nutrition. 2014; **2**:811-820

[54] Marco ML, Heeney D, Binda S, Cifelli CJ, Cotter PD, Foligné B. Health benefits of fermented foods: Microbiota and beyond. Current Opinion in Biotechnology. 2017;**44**:94-102. DOI: 10.1016/j.copbio.2016.11.010

[55] Brenton BP. Piki, polenta, and pellagra: Maise, nutrition, and nurturing the natural. In: Hosking R, editor. Nurture: Proceedings of the Oxford Symposium on Food and Cooking. Bristol, UK: Footwork; 2004. pp. 36-50

[56] Adavachi, Winfred M. "The role of fermented maize-based products on nutrition status and morbidity of children 6-59 months old in Western Kenya." Nairobi, Kenya: University of Nairobi; 2017. [PhD diss.]

[57] Bressani R. Protein quality of high-lysine maize for humans. Cereal Food World. 1991;**36**:806-811

[58] Anon. Nutritious Maize Boosts Growth of Children in Rural Ethiopia. Addis Ababa, Ethiopia: African Science News Service; 2008

[59] Naqvi S, Zhu C, Farre G, Ramessar K, Bassie L, Breitenbach J, et al. Transgenic multivitamin corn through biofortification of endosperm with three vitamins representing three distinct metabolic pathways. Proceedings of the National Academy of Sciences. 2009;**106**:7762-7767

[60] Li B, Liu H, Zhang Y, Kang T, Zhang L, Tong J. Constitutive expression of cell wall invertase genes increases grain yield and starch content in maize. Plant Biotechnology Journal. 2013;**11**:1080-1091

Chapter 3

Maize (*Zea mays*) Response to Abiotic Stress

Yousaf Ali, Taufiq Nawaz, Nazeer Ahmed, Muhammad Junaid, Mehwish Kanwal, Fazli Hameed, Saeed Ahmed, Rafi Ullah, Muhammad Shahab and Fazli Subhan

Abstract

The most extensively produced crop globally is Maize (*Zea mays*). Its response to diverse environmental stressors is dynamics and complicated, and it can be plastic (irreversible) or elastic (reversible). There is a wide range of soil and climatic conditions in which Maize can be grown. Climate change, for example, has the potential to impair grain quality and productivity of Maize all over the world. For the best harvest yield, the maize crop requires the right temperature. As a result of climate change, environmental stress factors such as abiotic and biotic stress factors are projected to intensify and become more common. Abiotic stress such as drought, temperature, and salinity are the major constraints limiting Maize's worldwide production (*Z. mays* L.). In places prone to various stresses, the development of stress-tolerant crop types will be useful. Drought, salinity, and temperature extremes are examples of abiotic factors that can significantly impact the development and growth of the plant. Furthermore, various management options available may aid in the development of strategies for better maize performance in abiotic stress conditions to understand the maize response to resistance mechanisms and abiotic stress. Therefore, this chapter will focus on the impact of abiotic stress regarding temperature on Maize.

Keywords: maize, drought, temperature, salinity

1. Introduction

The most important staple cereal crop grown for biofuel and food globally is Maize. After wheat and rice, it is 3rd significant crop grown [1, 2]. Studies have suggested that maize production must double, especially in developing nations, to meet the increasing animal and human consumption demand. The optimum temperature range responsible for higher maize production is 28–32°C, and it requires 500–800 mm of water to complete the life cycle [2].

Environmental conditions play a vital role in crop production. The yield and other characteristics of plants are determined by their genotypes and are highly influenced by environmental conditions. Under natural conditions, plants undergo different phases to complete their life cycle. In recent years, climatic parameters such as precipitation, temperature are being more unpredictable and resulted in prolonged drought, change in temperature beyond the optimal state. Such changes have challenged crop production. In the last two decades, crop productivity has improved. However, the susceptibility of plants to abiotic stress poses a new challenge to sustaining an increase in crop production with changing climatic patterns [3]. Abiotic stress-tolerant crops may be essential to maintain crop productivity in the future [4]. Plant cells activate signaling pathways that include plant hormones, transcription regulators, and signal transducers to respond to various stress. These multiple signals converge to regulate stress-inducible genes, producing proteins and enzymes for stress metabolism [5].

Maize, Wheat, Barley, canola, and other crops attacked by different insect pests which reduced their yield. These attack of insect pest may be due to some compound present in these crop which attract these compound [6–8]. However, these compounds also act as repellent. Insect pests prefer these crops for their progeny production and development to complete their life cycle [9, 10]. Several methods used to control the insect pest and increase crop yield. However, among these methods pesticides severally used in the world. Due to hazardous effect on the human health and environment alternative methods have been adopted to reduce to use of chemical and control insect pests [11–13].

As Maize is worldwide grown grown crops so its production is also threatened by moderate to severe droughts, high air temperature, and erratic rainfalls [14]. The major focus of maize research is to improve abiotic stress tolerance characters. However, it's challenging to identify genetic components responsible for abiotic stress tolerance [3]. Different complex quantitative traits potentially in correlation with other developmental characteristics are responsible for abiotic stress tolerance. These traits are governed by multiple quantitative trait loci (QTL) with small individual effects on the overall trait expression, making it more difficult to identify and modify [1]. This chapter aims to assess the impact of different abiotic stress, especially temperature, in maize production.

2. *Zea mays* and abiotic stress

The global drop in grain production of annual crops is accelerating as abiotic pressures, such as nutrient limits and drought, rise to the top of the constraint list [15]. Maize looks to be the most vulnerable crop regarding the effects of climate change on agriculture. Drought, severe heat, salt, and nutrient deficiency are all known to be key environmental factors that harm maize productivity worldwide [16]. Maize's yield and growth are badly affected by waterlogging, low or high temperatures, and intense droughts [17]. Furthermore, due to climate change, ambient temperatures are expected to alter, thereby altering drought frequencies and the intensity in various maize-growing regions globally [18]. Across most Indo-Gangetic plains and Sub-Saharan Africa, the variability of climatic conditions is responsible for nearly half (50%) of the total fluctuations in maize yields in these regions [19].

Abiotic stress in general, and drought in particular, are particularly harmful to maize yields, regardless of the germplasm and stress faced during a developmental stage of the plant [20]. According to some research, when temperatures rise in the world's major maize-producing regions, maturity times may shorten. In contrast, rising temperatures will alter metabolism, resulting in a loss in carbon uptake and, as a result, a decrease in pollination and grain set [21, 22]. Furthermore, high temperatures can cause plant moisture stress due to the soil's decreasing moisture content, in addition to broad-scale climatic variables altering rainfall patterns [23, 24]. Researchers discovered that from 1961 to 2002, the global production of Maize decreased by 8.3% with every degree Celsius increase in temperature, with part of the variation explained by variations in temperature, both minimum and maximum, and precipitation. As a result, even if Maize is given all of the necessary water, yields are expected to fall by 10–20% by the end of the twenty-first century due to severe climate change [25]. Simultaneously, the global agricultural sector must produce roughly 70% of food for a population expected to reach 9 billion or more by 2050 [26].

3. Drought stress

Drought is complex and destructive in plant biology to such an extent that it is compared with cancer in mammalian biology [27]. The effect of drought varies with the timing and intensity of stress on a plant's growth and development [28].

Maize is a drought-sensitive crop, mainly in a critical stage of growth such as the seedling stage and is grown in a wide range of climatic conditions from semi-arid to temperate regions, including drought-prone areas of Africa, North and South America, Asia and Europe [2]. Drought stress during vegetative growth, especially during V1 to V5, reduces plant growth, increases the vegetative growth period and reduces the growth period of the reproductive stage [29]. The relative water content and water potential are reduced under stressed conditions.

Plants undergo morphological and physiological changes under drought stress conditions. This process can be covered under three major categories. They are drought escape, drought avoidance and drought tolerance. The combined impact of these strategies is drought resistance [30]. According to Osmolovskaya et al. [30] drought resistance is the ability of plants to maintain favorable water balance and turgidity under water stress conditions. Drought escape is a strategy in which plant complete their life cycle before the onset of drought. They show seasonal responses [30, 31]. The drought avoidance strategy integrates increased water uptake and decreased water loss by plants. Plants develop strategies such as osmotic adjustment, an extension of antioxidant capacity, and desiccation tolerance to develop drought tolerance.

Further research has been conducted to identify drought-tolerant varieties. Along with advancements in technology, the research focus has changed from morphological characterization to identifying genes responsible for drought tolerance. Photosynthesis, a major metabolic pathway in plants, is sensitive to drought stress and is involved in plant response [32]. The photosynthetic pigments are damaged by drought, which decreases the light absorption efficiency of plants [2]. Though stomata closure is a way forward to ameliorate the adverse effects of drought, a decrease in stomata opening reduces the amount of CO_2 entering in leaves, which reduces carbon assimilation reaction and transpiration decreases root absorbance.

4. Salinity stress

Salinity stress is mostly caused by the high concentration of NaCl which induces abiotic stress in plants in irrigated and non-irrigated conditions. According to a global estimation, 20% of cultivated land and 50% of irrigated land is under salinity stress [33]. Salinity stress retards plant growth and productivity, mainly due to ion toxicity and osmotic stress. Thus, induced osmotic stress decreases stomata opening, reducing photosynthetic ability [34].

Besides limitation in photosynthetic ability, salinity stress causes degradation of enzymatic proteins in photosynthetic apparatus and chlorophyll degradation [34]. Furthermore, salinity stress causes secondary stress, in particular oxidative stress, mainly caused by ion toxicity and osmotic stress, which damage plant cells by excessive accumulation of Reactive Oxygen Species (ROS) [35]. ROS causes significant damage to proteins, nucleic acids, lipids, and photosynthetic pigments. As a result, antioxidant capacity and photosynthetic capacity are two important factors to consider in salinity stress studies [36].

Application of exogenous selenium (1 μM) alleviates inhibitory effects caused by salt stress. In an experiment, Jiang [36] studied different concentrations of Na_2SeO_3 (0, 1, 5 and 25 μM) on 15 days old maize plants. This study found that the application of 1 μM Se increases net photosynthetic rate, improves antioxidant defense mechanism and reduces chloroplast ultrastructure damage caused by NaCl.

5. Temperature stress

At leaf temperatures greater than 38°C, maize plants demonstrated a drop in net photosynthesis (Pn), and the decrease was particularly severe when the temperature was increased suddenly rather than slowly [37]. The reduction in photosynthesis was not due to stomata closure, as the transpiration rate rise in response to the increase in temperature. An increase in temperature greater than 32.5°C decreased the activation state of rubisco, which guided to complete inactivation at 45°C [38]. With the increase in leaf temperature, the level of 3-hosphoglyceric acid decreased. Rubisco activation acclimatized with increased leaf temperature and the acclimation process was associated with the expression of new activase polypeptide. Crafts-Brander and Salvucci concluded that the primary constraint responsible for the decrease in net photosynthesis at a temperature greater than 30°C was the inactivation of rubisco [38].

Maize is sensitive to chilling injury (below 15°C) and shows less adaptation growing in low temperatures [39]. Miedema found that 36% of the imbibed seeds died when exposed to 4°C for 28 days [40]. Sugar and amino acid exudation at lower temperatures may be linked to cell membrane failure. Young seedlings died after six days at 1°C and 8 days at 2.5°C. After 3 days of cooling, the Golgi bodies and inner mitochondrial membrane were destroyed, the endoplasmic reticulum was decreased, and lipid bodies accumulated.

Maize leaves are most sensitive to chilling injury. Chilling injury induces premature leaf senescence [39]. The combined exposure of Maize leaves to low temperature (10°C) and high light decreases CO_2 assimilation and leads to irreversible inhibition of photosynthesis [40].

Janda et al. [41] treated young maize seedlings grown in hydrophobic conditions with 0.5 mM salicylic acid, which protected plants in subsequent application of low-temperature stress. Salicylic acid pretreatment lowered catalase activity, which boosted antioxidant enzyme activity such as peroxidases and glutathione reductase, resulting in higher freezing tolerance in immature maize plants, according to Janda et al. [41]. Another research found a significant reduction in lipid peroxidation in Glycinebetaine (GB) cells compared to control during chilling [42]. This result implies that an increase in chilling tolerance may be caused by reducing lipid peroxidation of the cell membrane in the presence of GB.

Maize appears to have a harder time adjusting to low temperatures. This adaptation necessitates the capacity to germinate, grow, and mature at low temperatures and resistance to frost, chilling, and soil fungi during germination. Breeding for low-temperature adaptability has grown more essential as feed maize in northern areas has increased. Appropriate selection criteria are required for a sensible approach to this breeding task. As a result, we need to know first which plant characteristics limit maize output in a chilly climate, and second, how genetically variable those characteristics are.

5.1 Damage by low nonfreezing temperature

Temperatures below and near the germination and growth minimums in Maize can induce various physiological problems. Chilling damage is the medical term for these low-temperature effects. Chilling is not to blame for all of the negative impacts of cold weather. At temperatures above the chilling range, for example, low-temperature chlorosis occurs.

5.1.1 Chilling injury

The physiological damage induced by temperatures between 0 and 12°C is known as chilling injury [43]. Chilling is a problem for many thermophilic plants. The temperature and length of exposure determine the severity of the injury. Injury is usually not obvious during chilling, but appears once the temperature rises. Chilling injury causes wilting and browning of the leaves; severe chilling causes plants or plant sections to die. The chilling injury could be caused primarily by membrane dysfunction at low temperatures.

5.1.2 Chilling before emergence

Long-term cold temperatures destroy the Imbibed seeds that have been killed. Researcher also looked explored the effects of a 28-day cold treatment on six different types of plants. Varietal survival differences were observed. At 4 and 6°C, the average mortality was 36 and 21%, respectively; there was hardly little harm at 8 and 10°C. Sugars, and amino acids were exuded from maize seeds when incubated at a low temperature. This exudation was substantially higher at 6 than at lWC, and it could be linked to membrane malfunction at the lower temperature. They identified a specific chilling injury after imbibition in very dry seeds incubated at 5°C. During initial hydration, structural defects in the radicle caused the injured [44].

Young plants were injured by a 6-day exposure to 1°C and an 8-day exposure to 2.5°C. Chilling generated ultrastructural alterations in the meristematic cells of primary roots. The Golgi apparatus and inner mitochondria1 membranes were destroyed after 3 days of cooling, Lipid bodies had accumulated, and the endoplasmic reticulum had decreased. After 4 days of cooling, researchers discovered double cells made up of a small cell inside a larger cell. According to the findings, temperatures below roughly 6°C damage or kill immature maize plants. Temperature, length of cold treatment, developmental stage, and genotype all influence the severity of the injury [44].

In the field, there is no data on how cold affects germination. It is doubtful that chilling will have an impact on Maize's survival and emergence because the crop is sown late in the spring when soil temperatures rarely fall below 6°C for long periods.

5.1.3 Chilling after emergence

Using 7-day-old maize seedlings, Miedema [44] investigated chilling effects at 0.3°C and low light intensity. The seedlings were moved to a temperature of 21°C to explore the physiological and biochemical consequences. Leaf damage began to appear after 36 h of exposure to the cold, and by 72 h, the damage had become irreversible. Leaf extension at 21°C was drastically reduced after 24 and 36 h of chilling. Increasing ion leakage and oxygen uptake in chilled plant leaf segments due to uncoupling oxidative phosphorylation. Plants that were chilled for 72 h did not occur this.

According to the findings of several researchers, seedlings that had been chilled in an air-conditioned greenhouse at 2–4°C for 60 h developed transverse chlorotic bands the leaf blades 5–10 days after the chilling period. As a result of the chilling process, bands of color began to emerge on the blades used to generate the plant's curled appearance. The researchers discovered similar necrotic cross bands and other leaf damage in maize seedlings that had been exposed to 4°C in the dark for three days [44]. After transfer normal temperatures, the majority of the damage disappeared. After 6 days of exposure to 4°C, irreversible leaf damage occurred.

After 14 days of exposure to a daylight temperature of 10/4°C, chlorotic cross bands were developed, but not at 16/4°C. Chilling sensitivity was higher in the cell extension zone of the leaves than in full-grown or meristematic tissue. The tissue between the veins was chlorotic in some cross bands, whereas tissue along the veins (bundle sheath) was green. Various thermophilic Gramineae have chlorotic cross bands. Its found that transverse permanently chlorotic bands appeared in *Sorghum bicolor*, *Paspalum dilatatum*, and Digitaria smut-sii following a single cold night. Chlorophyll-deficient chloroplasts with disorderly lamellae were detected in most mesophyll cells in the chlorotic bands, whereas chloroplasts in bundle sheath cells were green and had a normal structure. The nucleus and mitochondria of chlorotic mesophyll cells showed no structural changes.

Because of the chilling sensitivity of the chloroplasts, leaves appear to be more sensitive to chilling than other organs. Chilling treatments cause visible injury to the roots of maize seedlings. In the case of Maize, there was just a minor amount of genetic heterogeneity in chilling-induced leaf damage [44]. Chlorotic cross bands are frequently seen after a cold spell in the field. They could result from a combination of low temperature and high light intensities, or they could be the effect of chilling during cold nights.

5.1.4 Chilling injury at high light levels

When maize leaves are exposed to a temperature of 10°C and a light intensity of 170 W m-*, they develop necrotic lesions, according to Taylor and Rowley [45]. With increasing exposure time to very low levels on the third day, the photosynthetic rate at 10°C steadily decreased. A permanent photosynthetic capacity reduction was caused by this chilling treatment; Photosynthesis at 25°C was reduced by 40% and 70% after 1.5 and 2.5 days of cooling. The chilling treatment did not affect chlorophyll levels. According to Taylor and Craig [46] in Sorghum, this form of injury was related to edoema and changes in the ultrastructure of chloroplasts. The membranes of the thylakoids first closed together as the starch grains vanished, but with more severe stress, the thylakoids moved apart, and granal stacking vanished. Paspalurn and soybean both had similar effects.

When maize plants were subjected to the light intensity of 13°C and 350 W mP2, a decrease in photosynthetic rate was seen, similar to what Taylor and Rowley [46] discovered. They found that throughout a 10-day exposure period, photosynthesis of maize seedlings cultured at 10°C and 105 Wm^{-2} fell only 30%. The photosynthetic rate immediately restored to its previous level when the seedlings were reintroduced to 22°C. Temperatures of around 10°C in combination with strong light levels, in general, cause the forms of damage already mentioned. When Taylor and Rowley [46] employed conditions similar to those experienced in the field at the start of the growing season, they discovered that seedling growth was impeded.

5.2 Male sterility induced by low-temperature

Plants grown in a greenhouse with short photoperiods and cool nights (10°C) showed male sterility throughout flowering. The tassel growth stage was used to gauge the degree of sterility achieved during the cold treatment. Low night temperatures have also been linked to male sterility in rice and Sorghum [47], but not in Maize under field circumstances.

6. Conclusion

Low temperatures (below 16°C) cause various physiological harm in maize seedlings. Plants are susceptible to chilling in the range of 0°C (or, more specifically, the freezing temperature of the tissue) to around 6°C. The injury severity depends on exposure duration and temperature. Low temperatures for short periods are not detrimental. With each stage of seedling development, the sensitivity to chilling increases. Membranes dysfunction has been linked to chilling injury. Young seedlings and Imbibed seeds exposed to low temperature are susceptible to some other, physiologically less defined, types of injury. Moreover, most types of cold injury report genetic variation; however, there were few interrelationship indications. This shows that different mechanisms cause the various types of low-temperature damage.

Acknowledgements

The authors are grateful to Dr. Saeed Ahmed (State University of Londrina, Brazil) for reviewing this chapter in the early stages.

Conflict of interest

The authors declare no conflict of interest.

Notes/thanks/other declarations

The authors are thankful to the agriculture department staff for their support and encouragement.

Author details

Yousaf Ali[1], Taufiq Nawaz[2], Nazeer Ahmed[3]*, Muhammad Junaid[4], Mehwish Kanwal[5], Fazli Hameed[3], Saeed Ahmed[6], Rafi Ullah[3], Muhammad Shahab[3] and Fazli Subhan[3]

1 Agronomy Department of Agriculture Extension, Mardan, Khyber Pakhtunkhwa, Pakistan

2 Department of Food Science and Technology, The University of Agriculture Peshawar, Khyber Pakhtunkhwa, Pakistan

3 Department of Agriculture, University of Swabi, Khyber Pakhtunkhwa, Pakistan

4 Graduate School of Life and Environmental Sciences, University of Tsukuba, Tsukuba, Ibaraki, Japan

5 Ministry of National Food Security and Research, Pakistan Tobacco Board, Pakistan

6 Department of Horticulture, Faculty of Agriculture, Gomal University, Dera Ismail Khan, Khyber Pakhtunkhwa, Pakistan

*Address all correspondence to: drnazeerento@gmail.com

References

[1] Miao Z, Han Z, Zhang T, Chen S, Ma C. A systems approach to a spatio-temporal understanding of the drought stress response in maize. Scientific Reports. Jul 26 2017;**7**(1):1-4

[2] Xie T, Gu W, Meng Y, Li J, Li L, Wang Y, et al. Exogenous DCPTA ameliorates simulated drought conditions by improving the growth and photosynthetic capacity of maize seedlings. Scientific Reports. Oct 4 2017;**7**(1):1-3

[3] Mao H, Wang H, Liu S, Li Z, Yang X, Yan J, et al. A transposable element in a NAC gene is associated with drought tolerance in maize seedlings. Nature Communications. Sep 21 2015;**6**(1):1-3

[4] Duvick DN. The contribution of breeding to yield advances in maize (*Zea mays* L.). Advances in Agronomy. Jan 1 2005;**86**:83-145

[5] Zandalinas SI, Mittler R, Balfagón D, Arbona V, Gómez-Cadenas A. Plant adaptations to the combination of drought and high temperatures. Physiologia Plantarum. Jan 2018;**162**(1): 2-12

[6] Ahmed N, Ahmad S, Ahmad S, Junaid M, Mehmood N. Population trend of aphid (*Lipaphis erysimi)* Kalt on canola (*Brassica* spp.) cultivar in Peshawar. Pakistan Entomology. 2013;**35**:135-138

[7] Ahmed N, Chamila Darshanee HL, Fu WY, Hu XS, Fan Y, Liu TX. Resistance of seven cabbage cultivars to green peach aphid (Hemiptera: Aphididae). Journal of Economic Entomology. Apr 2 2018;**111**(2):909-916

[8] Ahmed N, Darshanee HL, Khan IA, Zhang ZF, Liu TX. Host selection behavior of the green peach aphid, *Myzus persicae*, in response to volatile organic compounds and nitrogen contents of cabbage cultivars. Frontiers in Plant Science. Mar 12 2019;**10**:79

[9] Darshanee HL, Ren H, Ahmed N, Zhang ZF, Liu YH, Liu TX. Volatile-mediated attraction of greenhouse whitefly *Trialeurodes vaporariorum* to tomato and eggplant. Frontiers in Plant Science. Jul 20 2017;**8**:1285

[10] Saeed M, Ahmad T, Alam M, Al-Shuraym LA, Ahmed N, Alshehri MA, et al. Preference and performance of peach fruit fly (*Bactrocera zonata*) and Melon fruit fly (*Bactrocera cucurbitae*) under laboratory conditions. Saudi Journal of Biological Sciences. 2021 Dec 10. DOI: 10.1016/j.sjbs.2021.12.001

[11] Ahmed N, Huma Z, Rehman SU, Ullah M, Ahmed S. Effect of different plant extracts on termite species (*Heterotermis indicola*). Journal of Bioresource Management. 2016;**3**(2):2

[12] Ahmed N, Alam M, Saeed M, Ullah H, Iqbal T, Al-Mutairi KA, et al. Botanical insecticides are a non-toxic alternative to conventional pesticides in the control of insects and pests? In: Global Decline of Insects. 2021. DOI: 10.5772/intechopen.100416

[13] Iqbal T, Ahmed N, Shahjeer K, Ahmed S, Al-Mutairi KA, Khater HF, et al. Botanical insecticides and their potential as anti-insect/pests: Are they successful against insects and pests? In: Global Decline of Insects. IntechOpen; 2021. DOI: 10.5772/intechopen.100418

[14] Lobell DB, Roberts MJ, Schlenker W, Braun N, Little BB, Rejesus RM, et al. Greater sensitivity to drought accompanies maize yield increase in the US Midwest. Science. May 2 2014;**344**(6183):516-519

[15] Mueller ND, Gerber JS, Johnston M, Ray DK, Ramankutty N, Foley JA. Closing yield gaps through nutrient and water management. Nature. 2012 Oct;**490**(7419):254-257

[16] Tebaldi C, Lobell D. Differences, or lack thereof, in wheat and maize yields under three low-warming scenarios. Environmental Research Letters. May 24 2018;**13**(6):065001

[17] Ahuja I, de Vos RC, Bones AM, Hall RD. Plant molecular stress responses face climate change. Trends in Plant Science. Dec 1 2010;**15**(12):664-674

[18] Yang J, Sicher RC, Kim MS, Reddy VR. Carbon dioxide enrichment restrains the impact of drought on three maize hybrids differing in water stress tolerance in water stressed environments. International Journal of Plant Biology. 2014;**5**(1):5535. DOI: 10.4081/pb.2014.5535

[19] Ray DK, Gerber JS, MacDonald GK, West PC. Climate variation explains a third of global crop yield variability. Nature Communications. Jan 22 2015;**6**(1):1-9

[20] Bänzinger M. Breeding for drought and nitrogen stress tolerance in maize: From theory to practice. Cimmyt. 2000

[21] Iqbal MM, Arif M. Climate-change aspersions on food security of Pakistan. A Journal of Science for Development. 2010;**15**(1):15-23

[22] Moriondo M, Giannakopoulos C, Bindi M. Climate change impact assessment: The role of climate extremes in crop yield simulation. Climatic Change. 2011 Feb;**104**(3):679-701

[23] Lobell DB, Field CB. Global scale climate–crop yield relationships and the impacts of recent warming. Environmental Research Letters. 2007 Mar 16;**2**(1):014002

[24] Challinor AJ, Simelton ES, Fraser ED, Hemming D, Collins M. Increased crop failure due to climate change: Assessing adaptation options using models and socio-economic data for wheat in China. Environmental Research Letters. Sep 29 2010;**5**(3): 034012

[25] Xu H, Twine TE, Girvetz E. Climate change and maize yield in Iowa. PloS One. May 24 2016;**11**(5):e0156083

[26] Smith P, Gregory PJ. Climate change and sustainable food production. Proceedings of the Nutrition Society. Feb 2013;**72**(1):21-28

[27] Pennisi E. The blue revolution, drop by drop, gene by gene. Science. Apr 11 2008;**320**(5873):171-173

[28] Wang X, Wang H, Liu S, Ferjani A, Li J, Yan J, et al. Genetic variation in ZmVPP1 contributes to drought tolerance in maize seedlings. Nature Genetics. Oct 2016;**48**(10):1233-1241

[29] Aslam M, Maqbool MA, Cengiz R. Drought stress in maize (*Zea mays* l.) Effects, resistance mechanisms, global achievements and Springer $ briefs in Agriculture; 2015

[30] Osmolovskaya N, Shumilina J, Kim A, Didio A, Grishina T, Bilova T, et al. Methodology of drought stress research: Experimental setup and physiological characterization. International Journal of Molecular Sciences. Dec 2018;**19**(12):4089

[31] Basu S, Ramegowda V, Kumar A, Pereira A. Plant adaptation to drought stress. F1000Research. 2016;**5**

[32] Meng Q, Chen X, Lobell DB, Cui Z, Zhang Y, Yang H, et al. Growing sensitivity of maize to water scarcity under climate change. Scientific Reports. Jan 25 2016;**6**(1):1-7

[33] Wang Y, Gu W, Meng Y, Xie T, Li L, Li J, et al. γ-Aminobutyric acid imparts partial protection from salt stress injury to maize seedlings by improving photosynthesis and upregulating osmoprotectants and antioxidants. Scientific Reports. Mar 8 2017;**7**(1):1-3

[34] Munns R, Tester M. Mechanisms of salinity tolerance. Annual Review of Plant Biology. Jun 2 2008;**59**:651-681

[35] Chaves MM, Flexas J, Pinheiro C. Photosynthesis under drought and salt stress: Regulation mechanisms from whole plant to cell. Annals of Botany. Feb 1 2009;**103**(4):551-560

[36] Jiang C, Zu C, Lu D, Zheng Q, Shen J, Wang H, et al. Effect of exogenous selenium supply on photosynthesis, Na^+ accumulation and antioxidative capacity of maize (*Zea mays* L.) under salinity stress. Scientific Reports. Feb 7 2017;**7**(1):1-4

[37] Magar MM, Parajuli A, Shrestha J, Koirala KB, Dhital SP. Effect of PEG induced drought stress on germination and seedling traits of maize (*Zea mays* L.) lines. Türk Tarım ve Doğa Bilimleri Dergisi. 2019;**6**(2):196-205

[38] Crafts-Brandner SJ, Salvucci ME. Sensitivity of photosynthesis in a C4 plant, maize, to heat stress. Plant Physiology. Aug 1 2002;**129**(4):1773-1780

[39] Foyer CH, Vanacker H, Gomez LD, Harbinson J. Regulation of photosynthesis and antioxidant metabolism in maize leaves at optimal and chilling temperatures. Plant Physiology and Biochemistry. Jun 1 2002;**40**(6-8):659-668

[40] Miedema P. The effects of low temperature on *Zea mays*. Advance in Agronomy. 1982;**35**:93-128

[41] Janda T, Szalai G, Tari I, Paldi E. Hydroponic treatment with salicylic acid decreases the effects of chilling injury in maize (*Zea mays* L.) plants. Planta. Apr 1999;**208**(2):175-180

[42] Chen WP et al. Glycinebetaine increases chilling tolerance and reduces chilling-induced lipid peroxidation in *Zea mays* L. Plant Cell Environment. 2000;**23**(6):609-618

[43] Lyons JM. Annual Review of Plant Physiology. 1973;**24**:445-466

[44] Miedema P. The effects of low temperature on *Zea mays*. Advances in Agronomy. Jan 1 1982;**35**:93-128

[45] Taylor AO, Rowley J. Plants under climatic stress: I. Low temperature, high light effects on photosynthesis. Plant Physiology. May 1971;**47**(5):713-718

[46] Taylor AO, Craig AS. Plants under climatic stress: II. Low temperature, high light effects on chloroplast ultrastructure. Plant Physiology. May 1971;**47**(5):719-725

[47] Board JE, Peterson ML, Ng E. Floret sterility in rice in a cool environment 1. Agronomy Journal. May 1980;**72**(3): 483-487

Chapter 4

Climate-Smart Maize Breeding: The Potential of Arbuscular Mycorrhizal Symbiosis in Improving Yield, Biotic and Abiotic Stress Resistance, and Carbon and Nitrogen Sink Efficiency

Arfang Badji, Issa Diedhiou and Abdoulaye Fofana Fall

Abstract

Maize is part of the essential food security crops for which yields need to tremendously increase to support future population growth expectations with their accompanying food and feed demand. However, current yield increases trends are sub-optimal due to an array of biotic and abiotic factors that will be compounded by future negative climate scenarios and continued land degradations. These negative projections for maize yield call for re-orienting maize breeding to leverage the beneficial soil microbiota, among which arbuscular mycorrhizal fungi (AMS) hold enormous promises. In this chapter, we first review the components relevant to maize-AMF interaction, then present the benefits of arbuscular mycorrhizal symbiosis (AMS) to maize growth and yield in terms of biotic and abiotic stress tolerance and improvement of yield and yield components, and finally summarize pre-breeding information related to maize-AMF interaction and trait improvement avenues based on up-to-date molecular breeding technologies.

Keywords: maize, climate-smart breeding, arbuscular mycorrhizal fungi, pre-breeding and Breeding, sterss tolerance, GWAS, genomic selection

1. Introduction

By 2050, around 9.9 billion people will be living on earth, with an expected concurrent doubling of the global food demand; hence agricultural production must increase at the same rate [1, 2]. This exponential population growth comes with an expected doubling of meat consumption; hence, demand for cereals-based feeds such as maize (*Zea mays* L.) will follow the same trend [3]. Therefore, modern breeding programs must double current genetic gains [2, 4]. However, a 2013 study alerted that current yield increase trends of most staple crops, including maize, are insufficient to meet the 2050 food demands [5]. Several biotic and abiotic factors are determinant limitations to crop yield despite breeding efforts, and climate disturbances will further compound these yield-reducing stressors [2, 6].

With this background, it is apparent that crop improvement programs targeting only the inherent genetic makeups of plants to increase yields are not efficient enough to increase crop production sufficiently to meet future food and feed demands [4]. Furthermore, it is necessary to achieve the desired crop production without increasing production surfaces since land degradation will likely worsen shortly [3]. This alarming situation calls for re-orienting plant breeding programs to adopt new climate-smart breeding approaches to boost crop productivity sufficiently to levels that would match the current and expected population growth rates [4, 7]. Instead of just focusing on increasing crop yields through improving its inherent genetic makeup, plant breeders should explore the immediate growing environment of the crops they seek to improve.

One of the components of the immediate growing environments of crops is the soil microbiota composed of microorganisms including fungi, archaea, and bacteria that have co-evolved with plants, among which some form highly beneficial symbiotic relationships with plants helping them in nutrient uptake, growth regulation, biotic and abiotic resistance, which ultimately results in increased yields [8–10]. The soil microbiota is relevant since most plant traits of interest, including nutrient use efficiency, tolerance to drought, salt, pest, and diseases, and yield, are part of a system comprising complex plant-associated microorganisms [11]. Plants have developed the ability to modulate the composition and activity of their microbiota through the secretion molecules and other signaling compounds [12]. Plant breeders would greatly benefit from understanding the genetic basis of this interaction and its influence on traits of interest and using this knowledge during selection and stability studies [11].

Among these microorganisms, arbuscular mycorrhizal fungi (AMF) are significant elements of the soil–plant system and constitute 5 to 50% of the microbial biomass of soils (Olsson et al., 1999). Arbuscular mycorrhizal symbiosis (AMS) is the most widespread and oldest terrestrial symbiosis [13], formed by 80–90% of terrestrial vascular plants, including grasses such as maize [14]. Plants benefit from AMS better mineral nutrition, especially phosphorus uptake [15], which improves mycorrhized plants biomass compared to non-mycorrhizal plants [13, 16], through colonization of plant roots, production of large networks of extra-root mycelium in the soil, and aid in the uptake of mineral nutrients by hosts in exchange for carbohydrates [17]. Elements such as nitrogen, magnesium, calcium, potassium, and trace elements such as copper [18], zinc, or even iron, are better absorbed by the plant through the AMF symbiosis [19, 20].

Furthermore, AMF are also a significant component of soil fertility improvement [21], playing an essential role in the soil's physical, chemical, and biological components. They increase soil water holding capacity by improving its structure and inherent enzymatic activity by activating other microorganisms such as nitrogen-fixing bacteria [22]. AMF also participate in the nitrogen, phosphorus, and carbon cycles while correcting soil acidity [23, 24]. Regarding plant bioprotection, AMF play an essential role [25, 26] through, for instance, helping plants to thwart root-damaging nematodes [27, 28], pathogenic fungi such as verticillium wilt [29], and various pathogenic bacteria [30]. In addition, AMS provides plants with better resistance to abiotic stresses such as water and salt stress or the presence of heavy metals [31, 32].

Maize (*Z. mays* L.) is one of the essential staple food crops, therefore, essential to food security [33, 34]. Along with rice and wheat, it provides at least 30% of the food calories to more than 4.5 billion people in 100 countries [35, 36]. Maize is grown in more than 166 countries in the world, including tropical, subtropical, and temperate regions from mean sea level to 3000 m AMSL [37], and among cereals, it ranks highest in terms of grain yield per hectare worldwide [38]. Furthermore, besides its revenue generation as a cash crop, maize utilization is diversifying and joining new markets such as biofuel production and livestock feed [33, 34]. Therefore, maize productivity and production need to increase exponentially to support future food and feed demands as a food security crop. This increase will be required to occur under less inorganic input, severed land degradation, and aggravated climate disturbances that will profoundly disadvantage maize in its interaction with its biotic and abiotic environments. Considering the tremendous benefits of AMS to crops such as maize, we believe future maize research and breeding should adopt a more climate-smart approach by focusing more on understanding maize-AMF interaction and improving symbiotic capacity to boost yields. Therefore, in this chapter, we first review the components relevant to maize-AMF interaction, then present benefits of AMS in terms of biotic and abiotic stress tolerance and improvement of yield and yield components, and finally summarize pre-breeding information related to maize-AMF interaction and trait improvement avenues based on up-to-date molecular breeding technologies.

2. Arbuscular mycorrhizal fungi (AMF)

AMF are significant elements of the soil–plant system [13]. Mycorrhizae from the Greek "*myco*" for fungus and "*rhize*" for root essentially refers to the symbiotic association between fungi and plants' roots. AMF constitute 5 to 50% of the microbial biomass of soils. Mycorrhizal hyphae biomass can vary from 54 to 900 kg per hectare [39], or nearly 200 meters of hyphae per gram of soil [40]. AMS is the most widespread and oldest terrestrial symbiosis [41], formed by more than 80–90% of terrestrial vascular plants [14]. AMF inhabit all continents, from the subarctic islands to the Antarctic Peninsula [19, 42]. According to Wang et al. [9], AMF has co-evolved with plants for at least 400 million years, allowing plants the colonization of lands by plants through improved hydro-mineral nutrition.

AMF belong to the phylum of Glomeromycota [43], with a taxonomic classification of AMF originally based on morpho-anatomical observations of spores [44]. However, the advances in biomolecular tools have allowed the use of Polymerase Chain Reaction (PCR) in the classification of AMF through amplification of ribosomal regions (18S), which permits a better definition of species or even molecular taxa [45]. More than 250 species of Glomeromyceta are currently described, with a constantly updated taxonomy updated, resulting in new species and higher taxa being regularly introduced [46, 47]. AMF are obligate symbiont because of their inability to develop without a host plant [48], with a reproductive system that can be either clonal or asexual through spore or coenocyte formation [49]. Several studies have shown the existence of hyphal fusions, called anastomoses which lead to exchanges of nuclei and cytoplasm between species of the same genus [50–52], hence participating in the conservation of diversity and the complex genetics of AMF [53]. Some AMF species are homokaryotic, with identical nuclei in each spore; thus, the genetic variation is present in each spore, resulting in several different copies of the same gene [54]. Heterokaryotic species are characterized by different nuclei in each spore, resulting in the distribution of the genetic variation among the different nuclei inside a spore. In other words, the existence of several different genomes in each spore helps AMF to adapt to different environments [55]. Furthermore, there are four phases in the life cycle of AMF [56]: the spore germination and hyphae growth in the rhizosphere phase, the root infection phase by hyphae, stimulated by carbon dioxide (CO_2), and root exudates, which propagates in the root, the phase of root colonization with the formation of arbuscules and vesicles, and the phase of the development of external hyphae, resulting in an increased volume of soil explored by the roots and production of spores [57].

3. Survey of AMF that colonize maize

Host specificity of AMF is a longtime debate among researchers. Although many authors argued that AMF have no host-plant specificity [58], several studies tend to demonstrate the preference of some AMF genus to some plant species [59]. Crops like maize have a relatively high mycorrhizal dependency for plant growth and nutrient uptake [60]. Mycorrhizal dependency is an intrinsic property of every plant species that depend on the AMS. The type of crop, the soil properties, and the effect of the cropping system characterize mycorrhizal dependency [61]. There are three categories of plant species according to their mycorrhizal dependency. First, non-mycotrophic plants are capable of developing without the intervention of mycorrhizal fungi. Secondly, facultative mycotrophs require that reproduction occurs in the presence of mycorrhizal symbiosis only when the environment in which they grow is nutrient-limited. Lastly, obligate mycotrophs can only complete their development cycle associated with AMF. To study mycorrhizal dependency, the percentage of root colonized by AMF is a critical index. For example, the root colonization rate is 0% for *Brassica napus* and between 50 and 70% for *Z. mays* [62].

Furthermore, researchers use different methods such as trap cultures, wet sieving, morphological identification of spores, and molecular tools to determine AMF

AMF family	AMF species	Habitat	References
Acaulosporaceae	*Acaulospora sp*	Benin, Thailand	[63–65]
Acaulosporaceae	*Acaulospora longula*	Brazil	[66]
Acaulosporaceae	*Acaulospora rugosa*	Brazil	[66]
Acaulosporaceae	*Acaulospora scrobiculata*	Brazil	[66]
Acaulosporaceae	*Acaulospora morrowiae*	Brazil	[66]
Glomeraceae	*Glomus caledonium*	Japan	[67]
Entrophosporacae	*Entrophospora sp*	Thailand	[65]
Glomeraceae	*Glomus sp*	Benin, Hungary, Japan, Thailand	[63–65, 68–70]
Glomeraceae	*Glomus intraradices*	Germany, Switzerland	[71]
Glomeraceae	*Scutellospora sp*	Benin, Thailand	[63, 65]
Archaeosporaceae	*Archaeospora sp*	Hungary	[66, 69]
igasporaceae	*Gigaspora sp*	Benin, Hungary	[63, 64, 69]
Entrophosporaceae	*Entrophospora schenckii*	Thailand	[65]
Glomeraceae	*Glomus mosseae*	Germany, Switzerland	[65, 71]
Paraglomeraceae	*Paraglomus sp*	Hungary	[69, 71]
Gigasporaceae	*Scutellospora fulgida*	Thailand	[65]
Glomeraceae	*Glomus geosporum*	Thailand	[65]

Table 1.
Examples of AMF species associated with maize.

communities associated with maize. Based on such methods, many authors showed that maize crops could associate with different AMF species belonging to various genera (**Table 1**). *Paraglomeraceae, Aucolosporaceae, Gigasporaceae, Glomeraceae, Archaeosporaceae,* and *Paraglomeraceae* are the AMF families that associated with maize; however, the Glomus group preponderant [63, 64, 66, 68, 69, 71]. Although it is well established that the genus Glomus has the most widespread dispersion, maize genotypes and agricultural systems influence the mycorrhizal community. Evaluation of maize varieties with different genomes revealed colonization levels depend largely and continuously on maize genotypes within each germplasm [70]. Maize monoculture also reduces AMF diversity [68, 71, 72]. For instance, Archaeosporaceae and Paraglomeraceae group are not colonizers of maize grown in monoculture [69].

4. Phenotypic and molecular bases of maize-AMF symbiosis

Most terrestrial plants, including maize, interact with AMF under nutrient-limited conditions, mainly phosphorus. The physiological mechanisms underlying AMS establishment are under intense study using model plants, yet little information is currently available about molecular bases. Several studies have been performed for investigating the effect of AMF on gene expression through several approaches in different plant species over the last few years. Transcriptional analyses of few model plants like rice [73], Petunia [74], *Lotus japonicus* [75], *Medicago truncatula* [76], and tomato [77] allowed to identify genes involved in the AMS including genes encoding mycorrhizae-specific transporters. According to Willmann et al. [78], transporter genes are crucial for functional symbiosis. Many authors have defined four distinct stages of AMS based on the morphological analyses of the mutants.

The first step of an AMS is pre-contact signaling. It is characterized by a bi-directional exchange of signaling molecules and metabolic resources between AMF and plants. Indeed, plants produce strigolactone recognized by AMF, which exudes Myc-LCO, leading to deformation of absorbent hairs and nuclear calcium spiking [79]. The calcium oscillations are decoded by calcium and calmodulin-dependent protein kinases (CCaMK/DMI3), activating CYCLOPS/IPD3. These genes induce various micro RNA, transcription factors, and auxin signaling during AMS as documented by Diedhiou and Diouf [80]. Transcription factors regulate the signaling pathway during mycorrhization through interconnections, not yet clearly defined compared to Rhizobium/legume symbiosis.

Molecular recognition between partners is followed by contact between fungal hyphae and plant roots. This contact triggers a chain of events that starts with the hyphae branching, differentiating into hyphopodium or appressorium on the root surface. This structure prepares the penetration of the fungus into plant cells. Then, hydrolases and other molecules probably make the cell wall more flexible and cause the migration of the nucleus towards the appressorium, rearrangements of the cytoskeleton, and endoplasmic reticulum. These events lead to the subsequent formation of a pre-penetration apparatus (PPA) [57]. This apparatus facilitates the invasion of hyphae on epidermal and the first cortical cells [57]. According to some authors, PPA is responsible for forming a symbiotic interface and a new apoplastic compartment separating AMF and plant. During appressorium formation, defense genes are weakly activated in the plant [81]. Several genes such as VAPYRIN, NSP1/NSP2, and Cbf1/Cbf2 are involved in this step, as documented by Diedhiou and Diouf [80]. However, their precise function in root endosymbiosis remains unclear.

After physical contact between the two partners, hyphae fungal penetrate the cortical cells without damaging their plasma membrane, which invaginates and proliferates around the hyphae that develop inside these cells. This event results in forming an intra-radical mycelium. Intra-radical proliferation extends the colonization area to the intercellular space of the cortical parenchyma and inner cortical cells. Very few specific genes are involved in this step [80, 82]. This low expression supports the hypothesis that very few additional plants genes are activated after successful fungal colonization.

Intra-radical proliferation leads to the formation of arbuscules, which represent the final and most intimate step of the AMS. Indeed, intramatricial hyphae perforate the cell wall and penetrate inside the cell. They branch out to achieve a structure reminiscent of a small tree called an arbuscule which surrounds the cytoplasmic membrane to form the peri-arbuscular membrane (PAM). These modifications induce many changes in genes expression patterns even if their activity on the whole root level largely remains the same. Studies carried out on model plants allowed to identify mainly transcription factors and phosphorus transporters require to form arbuscules [21, 83]. The possible interconnections between these genes are described particularly between RAM1 and PT4/STR [84]. For maize, several analyses focused on identifying phosphate transporters whose Pht1;6 localized to arbuscule-containing cells [78]. It plays a criticssl role in the maintening the arbuscule function. Indeed, loss of function of Pht1;6 reduce root colonization with premature degeneration of the arbuscules. In addition, 13 Pi transporters were identified by Liu et al. [21]. Among them is ZmPt9 gene, which is different from members of the PHT1 gene family. Functional analysis indicates that ZmPt9 promotes the Pi transporter gene induction involved in Pi uptake [85]. Overexpression of ZmPt9 in Arabidopsis plant increases primary root length and lateral root formation. Furthermore, phosphorus content is higher in the transgenic plant compared to the wild type [86]. Recently, Wang et al. [87] showed that ZmPt7 regulates Pi acquisition, and its transport is mediated by phosphorylation.

5. Impact of AMF on maize resistance to biotic and abiotic stress

Maize is one of the essential sources of carbohydrate globally [88]; however, abiotic stresses and plant pests and diseases are significant threats in maize production worldwide, and future climate disturbances will further compound these scenarios [89]. AMS improves plant growth, hydro-mineral nutrition, and physiology under various environmental stress conditions like salinity, drought, and the presence of heavy metals [90], as well as resistance to biotic stresses such as pests, diseases, pathogen and weeds [91]. The benefits of AMF to plant partners vary depending on the type of stress [92].

AMF adapt to biotic and abiotic stresses independently of its host plant [14] and respond to stresses such as pests, diseases, pathogen, weeds, drought, extreme temperatures, salinity, and heavy metals [93–95]. Extensive evidence shows that AMF can control plant fungal, viral, and bacterial diseases (Himaya et al., 2021). The adaptation mechanisms of AMF to these biotic stresses are generally linked to pathogen resistance, including competition for colonization sites and improvement of the defense system of the plants [14]. Gerlach et al. [96] reported changes in leaf's elemental concentration, resource reallocation, especially for carbohydrates and amino acids, and expression of defense-related genes under maize-AMF symbiosis. Patanita et al. [97] demonstrated the benefits of mycorrhization in the control of *Magnaporthiopsis maydis* also called *Harpophora maydis* [98], the cause of late wilt disease of maize, which causes up to 50% grain yield losses in many countries

[99, 100]. In addition, *Fusarium* and *Aspergillus* are two of the most dominant fungal pest species of maize, causing acute diseases and yield losses and majorly responsible for deterioration and losses on maize plants [101, 102]. Olawuyi et al. [103] investigated the effect AMF on Aspergillus niger and revealed that *Glomus deserticola* was an effective biocontrol agent against *Aspergillus niger*, the soil-borne pathogen of maize. *Glomus clarum* and *Glomus deserticola* a have biocontrol potential against *Fusarium verticillioides* [104]. Downy mildew disease caused by *Peronosclerospora* is responsible for decreasing maize production (Soenartiningsih and Talanca 2010). The combination of botanical fungicides (Turmeric rhizome and betel leaves) with AMF (*Enthropospora* sp., *Gigaspora* sp., and *Glomus* sp.) and *Trichoderma asperellum* can reduce the incidence of downy mildew by extending the incubation period and increasing the dry weight of maize shoots [105]. *Striga hermonthica* is one of the most critical biotic constraints affecting maize crops in sub-Saharan Africa. The high infestation of this parasitic plant has forced many poor farmers to abandon their farms [106]. Several studies have demonstrated that AMF can inhibit or suppress Striga germination, especially on cereal crops such as maize [107, 108]. Othira [109] carried out a study that confirmed the effectiveness of AMF in protecting maize against Striga infestation, promoting crop growth, and reducing *Striga* plant incidence, plant biomass, and phosphate content. He evidenced that AMF (*Gigaspora margarita*) enhanced the performance of the maize plant host, allowing it to resist better *Striga* damage [109].

In addition, AMF also helps maize plants cope with abiotic stresses such as salinity, drought, extreme temperature, and heavy metal. Various mechanisms explain abiotic stress biological regulation through AMF, such as increased hydromineral nutrition, ion selectivity, gene regulation, production of osmolytes, and the synthesis of phytohormones and antioxidants [14]. For instance, *Rhizophagus irregularis*, an AMF species, can improve maize drought tolerance through enhancing apoplastic water flow [110]. According to Mathur and Jajoo [111], *Glomus Funneliformis* can help maize resist extreme temperatures by regulating the photosystem (PS) II heterogeneity. Studies carried out by Estrada et al. [94] demonstrated that AMF species such as *R. irregularis*, *Septoglomus constrictum,* and *Claroideoglomus etunicatum* improve maize tolerance to salinity. The authors showed that these AMF species improve K^+ and Na^+ homeostasis, shoot and root dry weights, shoot K concentration, and reduced Cl and Na contents in shoots. AMF can also play a role in maize tolerance to heavy metals. Indeed, maize inoculation with some Glomus isolates can improve maize dry weight and contents of essential elements (K, P, and Mg) [93].

Drought is also one of the significant stresses that can reduce maize productivity [112]. Water constraints decrease the photosynthetic activity of plants, which close their stomates to minimize water loss, decreasing productivity [113, 114]. Several studies demonstrated that AMF improves crops performance under drought stress [115]. Mycorrhizal maize deals with water deficit through drought mitigation and drought tolerance [116]. A drought mitigation strategy is mediated by indirect AMF benefits and enhanced water uptake. In contrast, drought tolerance involves a combination of direct AMF benefits that improve the innate ability of the plant to cope with stress [117]. Furthermore, inoculation with AMF improves strigolactone and auxin responses to drought stress Ruiz-Lozano et al. [118]. These two critical hormones in plant resilience to abiotic stress [119].

6. Impact of AMF on maize carbon and nitrogen sink efficiency

Carbon (C) and nitrogen (N) are indispensable mineral elements for plant growth and development. AMF plays a vital role in maintaining soil quality by increasing carbon mineralization. After inoculating by *Glomus etunicatum*, the contents of dissolved organic carbon (DOC), microbial biomass carbon (MBC), and readily oxidizable carbon (ROC) increase in the soil rhizosphere of maize [120]. *Rhizoglomus intraradices* colonization improves the active carbon pools such as water-soluble carbon, hot water-soluble carbon, biomass carbon up to 305 mg.kg − 1, and passive pools such as soil organic carbon up to 4.31 mg.g − 1 compared to the control [121].

Plants can uptake nitrogen from the soil in the form of organic or chemical fertilizers [122] or establish beneficial associations with microbes that facilitate plant N acquisition [123–126]. Microbes convert different forms of N that plants can use following chemical reactions carried out by living microorganisms such as bacteria, archaea, and fungi. Bacteria like Rhizobia and Frankia are the leading nitrogen suppliers to legumes and actinorhizal plants, respectively [124, 125, 127]. Symbiotic mycorrhizal associations can also enhance plant N acquisition through endomycorrhizae or ectomycorrhizae [128]. AMF mobilizes N in the surrounding rhizosphere and provides it to the host plant [129, 130]. Indeed, AMF develop interconnected structures such as arbuscules, intraradical and extraradical mycelium that allow the N uptake [131] through up-regulating genes coding for NO3– and NH4+ transporters, including AMT3.1 [132]. ATM3.1 is the primary driver of NH4+ transfer to the plant colonized by AMF.

Another way to enhance plant nutrition, particularly N uptake, is to develop tripartite associations with bacteria and mycorrhizal fungi, even if they are not well characterized yet [133]. Indeed, bacteria of the genus Paenibacillus have been identified inside Laccaria bicolor cells and can stimulate in vitro production of *R. irregularis* spore and mycorrhizal plant colonization by *Glomus mossea* [134–136]. This stimulating effect enhances fungal growth that could favor the establishment of more efficient fungal and N2-fixing symbioses. Nevertheless, the contributions of AMF to nitrogen acquisition are little be known, even intercropping system between maize and nitrogen-fixing plant. In an intercropping between maize and soybean, common mycorrhizal networks (CMNs) regulate Nitrogen allocation to plant roots [137]. Co-inoculation with AMF and rhizobia transferred more than 54% more nitrogen from soybean to maize than inoculation with AMF alone [137]. Furthermore, a recent study questioned the relevance of the chitin-like N source, an organic N source for the AMF, in the N supply to plants. Experiments showed that only *R. irregularis* hyphae can access a significant fraction (>20%) of the organic N supplied as chitin into a pot zone but not *Andropogon gerardii* roots, and this Ni was transferred to the plants within as little time as five weeks [138].

Overall, the presented evidence suggests that AMF significantly impacts N use efficiency by mycorrhizal/rhizobial plants, and carbon allocation is effective even with a cereal-legume cropping system. Understanding these mechanisms in a climate change context is critical for introducing symbiotic microorganisms as organic fertilizer in both croplands and forests while taking care of the ecosystem services rendered by microbial symbionts. Moreover, there are few evience that AMS could mitigate greenhouse gase emission in several cropping systems through diverse mechanisms [139–145], opening avenues for breeding of climate-considerate crop varieties. In that regard, AMS was reported to mitigate N2O emissions in several crop-soil systems such as maize [139], tomato [143, 144], rice [145], and grassland [142, 144].

7. Impact of AMF on maize yield and allied traits

Currently, AMF are critical organic components in cropping systems. Their interaction with crops increases yields by promoting plant growth and nutrition capacity [10]. Several experiments investigated the effects of AMF inoculation on maize yield. According to Cozzolino et al. [146], inoculation by *Rhizophagus irregularis* increases maize stalk and leaf dry weight and grain yields compared to non-inoculated plants. They also found that colonization of maize by *R. irregularis* increases available soil phosphorus (P) concentrations suggesting that inoculated roots mobilize more P and water than wild type. Recently, Assogba et al. [147] revealed mixed effects of Glomeraceae and Acaulosporaceae groups on the growth of maize seedlings under greenhouse conditions. Glomeraceae group improve significantly fresh above and underground biomass to 54.97% and 42.94%, respectively, and 55.23% for the leaf area compared to the control. Moreover, maximum plant heights and number of leaves were obtained with the Acaulosporaceae group, having 20.55% and 17.04%, respectively, compared to 11.77% for the control.

Studies were also conducted to reveal the effects of AMF on maize yield under abiotic stresses such as drought, salinity, heavy metals. Drought is one of the significant stresses that negatively affect maize yield [112]. AMF applications under water deficit improve the maize yield in different irrigation regimes. *Rhizophagus irregularis* enhances shoot dry weight (SDW) between 26 and 35% under drought conditions [148, 149]. Limited irrigation causes a two-fold decrease of the dry shoot weight (SDW) in AMF-maize plants as compared to non-AMF plants (17% vs. 37%, respectively) [149]. Furthermore, co-inoculation of *Funneliformis mosseae* and *Pseudomonas fluorescens* (phosphate solubilizing bacteria) on maize improves vegetative and reproductive traits, root colonization, grain yield under water deficit while preserving natural resources such as P stocks [150]. According to Celebi et al. [151], *R. irregularis* significantly improves agro-morphological parameters even in restricted irrigation conditions and increases leaf and stem ratios. Like for drought, AMF can increase resistance to salinity through several mechanisms, thus improving yield. Zhang et al. [152] reported that Trichoderma and Stachybotrys could promote maize growth in saline soil. Indigenous AMF improve maize growth in saline fields by significantly increasing biomass production and promoting leaf proline accumulation and a higher $K+/Na+$ ratio [21]. Besides, *Glomus tortuosum* remarkably ameliorates dry mass and leaf area and enhances photosynthetic capacity by improving chlorophyll content and efficiently allowing light energy utilization, gas exchange, and rubisco activity under salinity stress [153].

Furthermore, several studies indicate that AMF can facilitate the revegetation of heavy metal contaminated soils and improve yields. Inoculation of maize by *Claroideoglomus etunicatum* in soils spiked with Lanthanum (La) significantly enhanced dry shoot weight and increased K, P, Ca, and Mg content in maize shoots between 27.40 and 441.77% [154]. Also, *C. etunicatum* decreased shoot La concentration by 51.53% in maize while root La concentration increased by 30.45%.

8. Pre-breeding and breeding perspectives to maize-AMF symbiosis

One of the pivotal paths towards climate resilience and reversing the predicted negative impacts on food security is the adoption of climate-smart breeding approaches to design of high-yielding crops adapted to climate disturbances, such as increased abiotic and biotic stress [155–157]. Considering the tremendous positive impact of AMF symbiosis on maize yield, biotic and abiotic stress tolerance, Carbon and Nitrogen sink capacity, greenhouse gas mitigation [10, 14, 90, 91, 141, 158–161], it is crucial to improve the symbiotic response capacity in the crop as a contribution to food security in the face of the changing climate scenarios and continuous land degradation [117, 162]. Besides, beneficial microorganisms such as AMF are not just traits influencers but rather are part of the phenotypic expression of plants in an integrated system started from nutrient mobilization to resource allocation to different plant functions, including resistance to biotic and abiotic tolerance, and nutrient efficiency and accumulation of assimilates in plant reserve organs which translates into yield [11, 163]. The current scientific information is more inclined towards a non-antagonistic impact of modern breeding activities on the mycorrhizal symbiotic capacity of maize [70, 96, 164–166], indicating the feasibility of efficient incorporating AMF-maize collaboration into ongoing breeding programs [165, 167, 168]. Therefore, the necessary continuous increase in crop yield to meet future food demands requires breeding efforts to improve symbiosis between critical crops such as maize with AMF [11, 166, 168].

For breeding to be efficient and achieve the expected genetic gains and progress, it is necessary to generate a host of useful pre-breeding information and devise breeding strategies based on state-of-art technologies selected through careful consideration of the factors that control maize-AMF interactions. In the coming sections, we review the critical pre-breeding information generated so far regarding maize symbiotic response to and interaction with AMF and identify areas necessitating further research to support breeding activities. We also devise breeding strategies to improve maize varieties' symbiotic response and interaction to AMF based on the current knowledge of maize-AMF interactions.

8.1 Genetic diversity and inheritance of maize response to AMF symbiosis

The first and foremost step towards climate-smart crops is harnessing genetic diversity to allow the selection of superior material for breeding [169]. The maize genotype affects AMF abundance in the soil [170]. Maize has high genetic variance in response to AMF colonization [11, 70, 164, 170], indicating the possibility of selection in breeding. However, little information is available regarding the type of gene action controlling the maize response to AMP symbiosis, hence the need for more research in this area to judge opportunities for trait improvement through breeding. The high genetic diversity revealed by the few studies that investigated this pre-breeding characteristic of the maize-AMF interaction should be verified in several other genetic and environmental backgrounds to confirm promises of fast breeding progress. However, breeding progress might be hampered by the generally low to moderate heritability of maize responsiveness to AMF colonization [171].

Expectations of slow breeding progress are especially true for the traditional phenotypic selection, which strongly relies on phenotyping, increasing cost and time for budget-constrained breeding programs. Also, information related to the genetic control of maize-AMF interactions as to the proportion of additive and non-additive gene action involved in maize symbiotic response to AMF colonization [11]. Understanding the genetic control of any trait is crucial to designing effective breeding strategies for its improvement [172]. Therefore, besides the need to conduct more studies targeting maize levels of genetic diversity in response to AMF colonization, determining the preponderance of additive vs. dominance or epistasis genetic control on the trait needs to be elucidated better to inform future breeding strategies. New breeding approaches, especially those relying on advances in marker technologies such as marker-assisted selection (MAS) and genomic selection (GS), but also transgenic and genome editing (GE) techniques, could help to accelerate the improvement of maize responsiveness to AMF colonization and increase yield and stress tolerance [173–176].

8.2 Genetic architecture of maize response to AMF

One of the first steps into implementing molecular breeding for maize response to AMF to increase yield performance is identifying genetic polymorphisms controlling the final maize benefit from the symbiosis [177]. However, very few studies undertook to map genomic regions associated with maize response to AMF symbiosis [168, 171]. Kaeppler et al. [164] conducted the first quantitative trait loci (QTL) mapping for maize interaction with AMF in a population generated from a cross between B73 and Mo17. They identified one QTL controlling maize responsiveness to AMF, and such a low number of QTL was attributable to the low heritability of the trait in their study. Twenty years later, Ramírez-Flores et al. [161] undertook another study to identify QTL that determined maize benefits from AMF symbiosis. Several QTL were identified in this study, suggesting a polygenic nature in the control of the trait, contrary to the monogenetic direction indicated by the earlier study [168, 171]. Considering the molecular complexity involved in the symbiosis process, from the recognition between AMF and plant to the effective establishment of the symbiotic relationship [158], the polygenic nature of Maize-AMF interaction is more plausible. However, more studies are required to confirm this hypothesis further.

Confirming the genetic architecture of maize-AMF interaction is pivotal since effective conventional or molecular breeding strategy design will depend on the number, siege of QTL controlling the traits and their interactions [178, 179]. These studies should be conducted in a wide variety of germplasm and geographical backgrounds to discover a comprehensive number of QTL that could accurately determine the genetic architecture of the trait through meta-analyses and other integrative studies [180].

8.3 Research perspectives and breeding strategies for improved maize response to AMS

Plant breeders generally are biased towards direct phenotypes, ignoring that most of these traits are mediated by beneficial microorganisms [11, 163, 167]. Although evidence points more towards a positive co-existence between modern plant breeding activities and practices, it is necessary to ascertain this status on target environments and maize populations. It is evident that response to AMF colonization is dependent on available resources such as soil phosphorus, crop species, and genotype [181]. The quantity and quality of soil phosphorus available to a particular crop are parameters that determine the maintenance of the diversity and quantity of the AMF community and their symbiotic capacity with maize [182]. Reports exist about the possibility of inhibition of the symbiosis between maize and AMF after artificial fertilization through the addition of external Phosphorus [183], making it necessary to adapt crop improvement for symbiotic capacity to target environments and cropping systems. **Figure 1** shows the cascade of pre-breeding and breeding activities that should be involved in a strategic crop improvement program targeting improved maize response to AMF colonization and symbiosis.

A typical breeding program for any trait should identify adequate germplasm, possibly including wild relatives, exotic accessions, and landraces, as a base population for breeding through careful mating designs and accelerate genetic gains towards possible variety release [184, 185]. The base population should be both phenotyped for target traits and genotyped with molecular markers to allow measuring phenotype and marker-based genetic diversity and population structure to optimize downstream research and breeding activities such as parent selection and cross designs for pre-breeding activities such as inheritance and genetic control studies, QTL mapping, and selection techniques such as phenotypic selection (PS), MAS, GS [186–189]. Phenotypic selection is a group of breeding methods basing the selection of superior genotypes for the next generation of for variety release on their observed phenotypic values. Phenotypic selection is best for highly

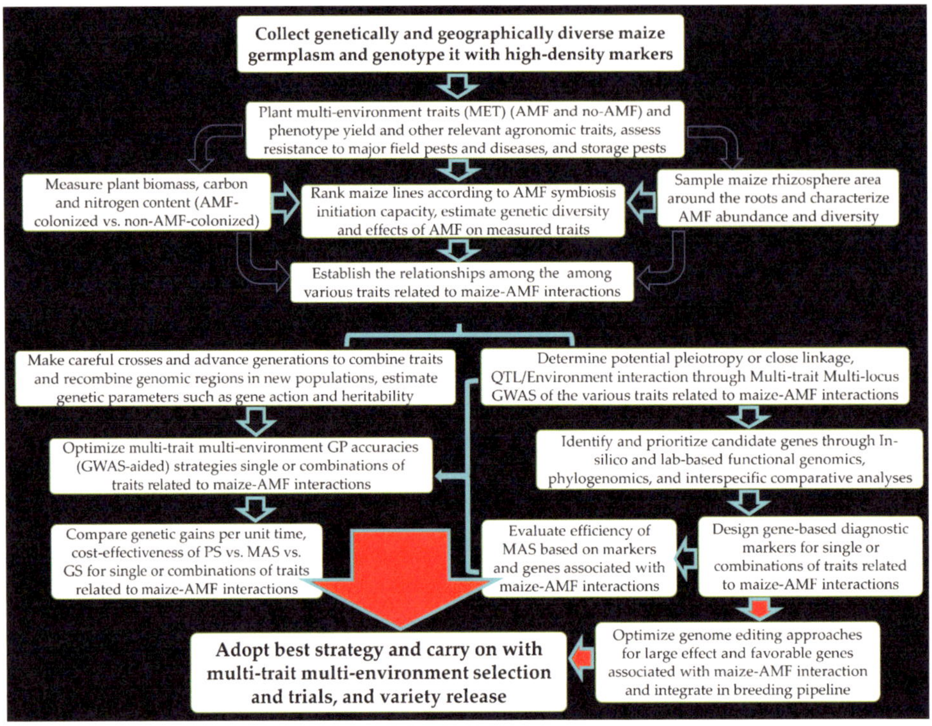

Figure 1.
Pre-breeding and breeding pipeline for improved maize-AMF symbiosis.

heritable and easy-to-measure traits. Both MAS and GS are based on selecting genotypes using molecular markers, albeit they are essentially different. MAS relies on mapped QTLs for a particular trait, of which it uses associated markers to select desired phenotypically unobserved lines. MAS works best when the trait is monogenic or oligogenic (controlled by one or a few large-effect QTL) [178, 190]. GS uses whole-genome markers to compute genomic-estimated breeding values of phenotypically unobserved genotypes as a basis for selection. GS performs best on polygenic traits that are controlled by multiple small-effect QTL, which characterizes most traits that plant breeders investigate [191–193].

In the case of maize interaction with AMF, phenotyping should be done with a control (non-mycorrhized plants) experiment to allow direct estimation of benefits offered by AMF symbiosis on traits of interest as in several studies [109, 194–196]. A comprehensive number of phenotypic, biochemical, and omic traits should be selected for phenotyping based on their direct or indirect involvement in or them being influenced by maize-AMF interactions to run univariate and multivariate analyses for genotype ranking, estimation of AMS effect on target traits, and strength and direction of relationships among traits [197]. Where possible, high-throughput phenotyping (HTP) techniques should be used to precisely measure and allow the accurate estimation of genetic and genomic parameters, including genetic control and inheritance, marker-trait association, and genomic prediction accuracies [198–200]. During the last decade, HTP technologies served to precisely measure the shoot biomass of tomato, barley [201], and Medicago [201, 202] growth trends under AMF colonization and to estimate nitrogen use efficiency of tomato, barley, and Medicago plants [203].

It is noteworthy that the inherent low allele diversity and low recombination rates arising from the bi-parental nature of such mapping populations and the short timespan from their generation to advanced stages used for mapping are critical limits to most of these studies based on genetic linkage based QTL mapping methods [204–207]. The low statistical power is because all the genetic and allele diversity only comes from the two parents crossed to generate the mapping population. The low resolution of QTL is caused by the short time for creating such populations, which, even with recombinant inbred lines, is still too little to allow enough recombination in the genome of the lines [206, 207]. These limitations lead to low statistical power for QTL discovery and low resolution of the genomic regions mapped [206, 207]. These shortcomings could have partly explained the low QTL number mapped by Kaeppler et al. [164], and that results from Ramírez-Flores et al. [161] might not have comprehensively captured the genetic architecture of maize-AF interaction. Genome-wide association studies (GWAS) is an alternative and complementary technique to pipe rental population-based QTL mapping from which it differs by the reliance on populations composed of diverse lines with historical recombination events. Consideration of genome-wide association studies (GWAS) should allow complementing traditional bi-parental QTL analyses, especially in Joint Linkage Association Mapping (JLAM), a technique that combines the strengths of both GWAS and biparental QTL mapping to alleviate their respective weaknesses [205]. Also, GWAS will increase the statistical power and resolution of the resulting QTL [204, 205, 208].

One of the main challenges breeders face is combining several traits of interest in elite lines due to the pervasive pleiotropic effect and close-linkage of genes controlling these traits, two genetic phenomena that yield similar phenotypic outcomes but are difficult to distinguish between each other unless specific analyses are performed [185, 209]. A prerequisite for efficiently achieving multiple-trait selection is delineating the genetic basis of the correlations among traits through multivariate analyses [210]. Several multivariate GWAS and GS exist in the perspective of genomics-aided multi-trait selection for maize response to AMF symbiosis. Multivariate methods allow leveraging shared genetic information among traits and possibly environments to increase statistical power and accuracy [210, 211]. Also, GWAS could complement GS by including GWAS-discovered QTL as fixed effects in GS models, which is reported to improve prediction accuracy, thereby increasing genetic gains per unit time [173].

Since its invention, GWAS has evolved, moving from single-locus single-trait mixed linear models proposed by Yu et al. [212] to multi-locus multi-traits algorithms, which, unlike the former, jointly test associations between several traits and all genome-wide markers [213, 214]. Single-trait mixed linear models suffer from several weaknesses, including high rates of false-negative associations caused by multiple testing issues that require stringent Bonferroni thresholds [215]. In contrast, multi-locus multi-traits algorithms have better statistical power by avoiding correcting for multiple testing [214, 216]. However, these methods are still inefficient in differentiating between the two causes of trait correlations [216]; instead, integration of structural equation modeling (SEM) to GWAS is necessary [217]. Several GWAS packages that incorporate SEM are available for use in the case of maize-AMF interactions, for instance, GW-SEM [218], SEM-GWAS [219], GenomicSEM [220]. A more advanced software package is the multi-trait multi-locus Structural Equation Modeling (mtmlSEM) that considers, besides the multi-trait framework, a multi-locus approach to model associations between multiple traits and all loci simultaneously using SEM [221]. Also, GWAS results should be complemented with a robust candidate gene discovery and In Silico and lad-based prioritization steps to allow selection of high-confidence trait-associated genes that could be used in molecular breeding techniques such as MAS GE [222–225]. GE is a novel molecular breeding technique that, after mapping a genome region with an unfavorable genetic effect or with the potential of improving a trait, is used to precisely modify, insert, replace, or delete DNA in a genome [226].

Determining the genetic architecture of maize-AMF interactions will allow breeders to decide what selection approach would yield better genetic gains in a shorter time with a competitive budget requirement. However, considering the complexity of the molecular basis of symbiosis (see Section 7 of this chapter) and the probable polygenic nature of the phenomenon [168], it is expected that PS or MAS might not be efficient [227–230]. GS, especially combined with HTP technologies, should accelerate genetic gains while reducing overall variety development costs [231]. For complex and polygenic traits such as maize-AMF interactions subject to several non-genetic influences, multi-trait GS models, especially those considering multi-environment trials (MET) such as R packages BMTME [232], would be of tremendous benefit. Multi-trait multi-environment GS methods are being routinely used for diverse traits of diverse crops, including maize [233–236].

Acknowledgements

The Authors thank the Carnegie Corporation of New York for funding the Post-Doctoral Fellowship of A.B. through the Regional Universities Forum for Capacity Building in Agriculture (RUFORUM), Grant number: RU-NARO/2020/Post-Doc/02. A.F.F is supported by the Regional Academic Exchanged for Enhanced Skills in Fragile Ecosystem Management (REFORM) Program. I.D is supported by DST-FICCI sponsored by CV Raman International Fellowship for African Researchers.

Conflict of interest

The authors declare no conflict of interest.

Author details

Arfang Badji[1,2*], Issa Diedhiou[3,4†] and Abdoulaye Fofana Fall[3,5†]

1 National Crops Resources Research Institute (NaCRRI), Kampala, Uganda

2 Department of Agricultural Production, Makerere University, Kampala, Uganda

3 Département de Biologie Végétale, Université Cheikh Anta DIOP de Dakar of Dakar, Dakar-Fann, Senegal

4 Ecole Supérieure de Génie Industriel et Biologique, Dakar-Fann, Senegal

5 Faculty of agriculture, African Center of Excellence in Agroecology and Livelihood Systems, Uganda Martyrs University, Kampala, Uganda

*Address all correspondence to: arbad2009@live.fr

† These authors contributed equally to this chapter.

References

[1] Islam SMF, Karim Z. World's Demand for Food and Water: The Consequences of Climate Change. In: Desalination - Challenges and Opportunities [Internet]. IntechOpen; 2020. p. 13. Available from: https://www.intechopen.com/books/advanced-biometric-technologies/liveness-detection-in-biometrics

[2] Myers SS, Smith MR, Guth S, Golden CD, Vaitla B, Mueller ND, et al. Climate Change and Global Food Systems: Potential Impacts on Food Security and Undernutrition. Annu Rev Public Health. 2017;38:259-77.

[3] Raimondo M, Nazzaro C, Marotta G, Caracciolo F. Land degradation and climate change: Global impact on wheat yields. L Degrad Dev. 2021;32(1):387-98.

[4] Voss-Fels KP, Stahl A, Hickey LT. Q&A: Modern crop breeding for future food security. BMC Biol. 2019;17(1):1-7.

[5] Ray DK, Mueller ND, West PC, Foley JA. Yield Trends Are Insufficient to Double Global Crop Production by 2050. PLoS One. 2013;8(6).

[6] Raza A, Razzaq A, Mehmood S, Zou X, Zhang X, Lv Y, et al. Impact of Climate Change on Crops Adaptation and Strategies to Tackle Its Outcome: A Review. Plants [Internet]. 2019;8(2):34. Available from: http://www.mdpi.com/2223-7747/8/2/34

[7] Mba C, Ghosh K, Guimaraes EP. Re-orienting crop improvement for the changing climatic conditions of the 21st century [electronic resource]. Agric food Secur [Internet]. 2012;1(1):6. Available from: http://dx.doi.org/10.1186/2048-7010-1-7%5Cnhttp://search.ebscohost.com/login.aspx?direct=true&db=agr&AN=IND44875826&site=ehost-live

[8] Compant S, Samad A, Faist H, Sessitsch A. A review on the plant microbiome: Ecology, functions, and emerging trends in microbial application. J Adv Res. 2019;19:29-37.

[9] Wang B, Yeun LH, Xue J, Liu Y, Ané J, Qiu Y. Presence of three mycorrhizal genes in the common ancestor of land plants suggests a key role of mycorrhizas in the colonization of land by plants. New Phytol [Internet]. 2010 Apr 6;186(2):514-25. Available from: https://onlinelibrary.wiley.com/doi/10.1111/j.1469-8137.2009.03137.x

[10] Real-Santillán RO, del-Val E, Cruz-Ortega R, Contreras-Cornejo HÁ, González-Esquivel CE, Larsen J. Increased maize growth and P uptake promoted by arbuscular mycorrhizal fungi coincide with higher foliar herbivory and larval biomass of the Fall Armyworm *Spodoptera frugiperda*. Mycorrhiza. 2019;29(6):615-22.

[11] Hohmann P, Messmer MM. Breeding for mycorrhizal symbiosis: focus on disease resistance. Euphytica. 2017; 213(5).

[12] Mendes R, Garbeva P, Raaijmakers JM. The rhizosphere microbiome: Significance of plant beneficial, plant pathogenic, and human pathogenic microorganisms. FEMS Microbiol Rev. 2013;37(5):634-63.

[13] Duponnois R, Ouahmane L, Kane A, Thioulouse J, Hafidi M, Boumezzough A, et al. Nurse shrubs increased the early growth of Cupressus seedlings by enhancing belowground mutualism and soil microbial activity. Soil Biol Biochem. 2011;43(10):2160-8.

[14] Diagne N, Ngom M, Djighaly PI, Fall D, Hocher V, Svistoonoff S. Roles of arbuscular mycorrhizal fungi on plant growth and performance: importance in biotic and abiotic stressed regulation. Diversity. 2020;12(10):1-25.

[15] Smith SE, Jakobsen I, Grønlund M, Smith FA. Roles of arbuscular mycorrhizas in plant phosphorus nutrition: Interactions between pathways of phosphorus uptake in arbuscular mycorrhizal roots have important implications for understanding and manipulating plant phosphorus acquisition. Plant Physiol. 2011;156(3):1050-7.

[16] Egerton-Warburton LM, Querejeta JI, Allen MF. Common mycorrhizal networks provide a potential pathway for the transfer of hydraulically lifted water between plants. J Exp Bot. 2007;58(6):1473-83.

[17] Alvarado-López CJ, Dasgupta-Schubert N, Ambriz JE, Arteaga-Velazquez JC, Villegas JA. Lead uptake by the symbiotic *Daucus carota* L.–Glomus intraradices system and its effect on the morphology of extra- and intraradical fungal microstructures. Environ Sci Pollut Res. 2019;26(1): 381-91.

[18] Facelli E, Smith SE, Smith FA. Mycorrhizal symbiosis overview and new insights into roles of arbuscular mycorrhizas in agro- and natural ecosystems. Australas Plant Pathol. 2009;38(4):338-44.

[19] Smith SE, Read D. Mycorrhizal Symbiosis. Third. Vol. 137, Soil Science. 2010. 204 p.

[20] Ercoli L, Schüßler A, Arduini I, Pellegrino E. Strong increase of durum wheat iron and zinc content by field-inoculation with arbuscular mycorrhizal fungi at different soil nitrogen availabilities. Plant Soil. 2017;419(1-2): 153-67.

[21] Liu F, Xu Y, Jiang H, Jiang C, Du Y, Gong C, et al. Systematic identification, evolution and expression analysis of the *Zea mays* PHT1 gene family reveals several new members involved in root colonization by arbuscular mycorrhizal fungi. Int J Mol Sci. 2016;17(6):1-18.

[22] Sadhana B. Review Article Arbuscular Mycorrhizal Fungi (AMF) as a Biofertilizer- a Review. IntJCurrMicrobiolAppSci [Internet]. 2014;3(4):384-400. Available from: http://www.ijcmas.com

[23] Jamiołkowska A, Księzniak A, Gałązka A, Hetman B, Kopacki M, Skwaryło-Bednarz B. Impact of abiotic factors on development of the community of arbuscular mycorrhizal fungi in the soil: A Review. Int Agrophysics. 2018;32(1):133-40.

[24] Parihar M, Rakshit A, Meena VS, Gupta VK, Rana K, Choudhary M, et al. The potential of arbuscular mycorrhizal fungi in C cycling: a review. Arch Microbiol [Internet]. 2020;202(7):1581-96. Available from: https://doi.org/10.1007/s00203-020-01915-x

[25] Eke P, Chatue Chatue G, Wakam LN, Kouipou RMT, Fokou PVT, Boyom FF. Mycorrhiza consortia suppress the fusarium root rot (Fusarium solani f. sp. Phaseoli) in common bean (*Phaseolus vulgaris* L.). Biol Control [Internet]. 2016;103:240-50. Available from: http://dx.doi.org/10.1016/j.biocontrol.2016.10.001

[26] Smms H, Sivasubramaniam N, Smms A. A Review on Role of Mycorrhizal Fungi in Plant Disease Management. 2021;41-50.

[27] Ceustermans A, Van Hemelrijck W, Van Campenhout J, Bylemans D. Effect of arbuscular mycorrhizal fungi on pratylenchus penetrans infestation in apple seedlings under greenhouse conditions. Pathogens. 2018;7(4).

[28] Schouteden N, Waele D De, Panis B, Vos CM. Arbuscular mycorrhizal fungi for the biocontrol of plant-parasitic nematodes: A review of the mechanisms involved. Front Microbiol. 2015; 6(NOV):1-12.

[29] Zhang W, Zhao F, Jiang L, Chen C, Wu L, Liu Z. Different Pathogen

Defense Strategies in Arabidopsis: More than Pathogen Recognition. Cells [Internet]. 2018 Dec 7;7(12):252. Available from: http://www.mdpi.com/2073-4409/7/12/252

[30] Singh I, Giri B. Arbuscular mycorrhiza mediated control of plant pathogens. Mycorrhiza - Nutr Uptake, Biocontrol, Ecorestoration Fourth Ed. 2018;131-60.

[31] Abd-Alla MH, Nafady NA, Bashandy SR, Hassan AA. Mitigation of effect of salt stress on the nodulation, nitrogen fixation and growth of chickpea (*Cicer arietinum* L.) by triple microbial inoculation. Rhizosphere [Internet]. 2019;10(January):100148. Available from: https://doi.org/10.1016/j.rhisph.2019.100148

[32] Bothe H. Arbuscular mycorrhiza and salt tolerance of plants. Symbiosis. 2012;58(1-3):7-16.

[33] Renzaho AMN, Kamara JK, Toole M. Biofuel production and its impact on food security in low and middle income countries: Implications for the post-2015 sustainable development goals. Renew Sustain Energy Rev [Internet]. 2017;78(May 2016):503-16. Available from: http://dx.doi.org/10.1016/j.rser.2017.04.072

[34] James A, Zikankuba VL. Mycotoxins contamination in maize alarms food safety in sub-Sahara Africa. Food Control [Internet]. 2018;90:372-81. Available from: http://linkinghub.elsevier.com/retrieve/pii/S0956713518301257

[35] Shiferaw B, Prasanna BM, Hellin J, Bänziger M. Crops that feed the world 6. Past successes and future challenges to the role played by maize in global food security. Food Secur. 2011;3(3):307-27.

[36] Cairns JE, Hellin J, Sonder K, Araus JL, MacRobert JF, Thierfelder C, et al. Adapting maize production to climate change in sub-Saharan Africa. Food Secur. 2013;5(3):345-60.

[37] Kumar P, Singh R, Jaswinder SBS, Sekhar KJC, Soujanya PL. An overview of crop loss assessment in maize. 2018;(August 2019).

[38] Amissah S, Osekre EA, Nyadanu D, Akromah R, Afun JVK, Adu Amoah R, et al. Inheritance and combining ability studies on grain yield and resistance to maize weevil (*sitophilus zeamais*, motschulsky) among extra early quality protein maize inbred lines. Ecol Genet Genomics [Internet]. 2019 Oct;12:100043. Available from: https://doi.org/10.1016/j.egg.2019.100043

[39] Zhu Y-G, Miller MR. Carbon cycling by arbuscular mycorrhizal fungi in soil–plant systems. Trends Plant Sci [Internet]. 2003 Sep;8(9):407-9. Available from: https://linkinghub.elsevier.com/retrieve/pii/S1360138503001845

[40] Leake J, Johnson D, Donnelly D, Muckle G, Boddy L, Read D. Networks of power and influence: The role of mycorrhizal mycelium in controlling plant communities and agroecosystem functioning. Can J Bot. 2004;82(8):1016-45.

[41] Fitter A, Gilligan C, Hollingworth K, Kleczkowski A, Twyman R, Pitchford J. Biodiversity and ecosystem function in soil. Funct Ecol. 2005;369-77.

[42] Newsham KK, Upson R, Read DJ. Mycorrhizas and dark septate root endophytes in polar regions. Fungal Ecol [Internet]. 2009;2(1):10-20. Available from: http://dx.doi.org/10.1016/j.funeco.2008.10.005

[43] Stürmer SL. A history of the taxonomy and systematics of arbuscular mycorrhizal fungi belonging to the phylum Glomeromycota. Mycorrhiza. 2012;22(4):247-58.

[44] Morton JB, Benny GL. Revised classification of arbuscular mycorrhizal fungi (Zygomycetes): a new order, Glomales, two new suborders, Glomineae and Gigasporineae, and two new families, Acaulosporaceae and Gigasporaceae, with an emendation of Glomaceae. Mycotaxon (USA). 1990;

[45] Young JPW. A molecular guide to the taxonomy of arbuscular mycorrhizal fungi. New Phytol. 2012;193(4):823-6.

[46] Krüger M, Krüger C, Walker C, Stockinger H, Schüßler A. Phylogenetic reference data for systematics and phylotaxonomy of arbuscular mycorrhizal fungi from phylum to species level. New Phytol. 2012;193(4): 970-84.

[47] Redecker D, Schüßler A, Stockinger H, Stürmer SL, Morton JB, Walker C. An evidence-based consensus for the classification of arbuscular mycorrhizal fungi (Glomeromycota). Mycorrhiza. 2013;23(7):515-31.

[48] Berruti A, Lumini E, Balestrini R, Bianciotto V. Arbuscular mycorrhizal fungi as natural biofertilizers: Let's benefit from past successes. Front Microbiol. 2016;6(JAN):1-13.

[49] Kuhn G, Hijri M, Sanders IR. Evidence for the evolution of multiple genomes in arbuscular mycorrhizal fungi. Nature. 2001;414(6865):745-8.

[50] Giovannetti M. Structure, extent and functional significance of belowground arbuscular mycorrhizal networks. Mycorrhiza State Art, Genet Mol Biol Eco-Function, Biotechnol Eco-Physiology, Struct Syst (Third Ed. 2008;59-72.

[51] De La Providencia IE, De Souza FA, Fernández F, Delmas NS, Declerck S. Arbuscular mycorrhizal fungi reveal distinct patterns of anastomosis formation and hyphal healing mechanisms between different phylogenic groups. New Phytol. 2005;165(1):261-71.

[52] Chagnon PL. Ecological and evolutionary implications of hyphal anastomosis in arbuscular mycorrhizal fungi. FEMS Microbiol Ecol. 2014;88(3): 437-44.

[53] Cárdenas-Flores A, Draye X, Bivort C, Cranenbrouck S, Declerck S. Impact of multispores in vitro subcultivation of Glomus sp. MUCL 43194 (DAOM 197198) on vegetative compatibility and genetic diversity detected by AFLP. Mycorrhiza. 2010;20(6):415-25.

[54] Hijri M, Sanders IR. Low gene copy number shows that arbuscular mycorrhizal fungi inherit genetically different nuclei. Nature. 2005;433 (7022):160-3.

[55] Croll D, Sanders IR. Recombination in Glomus intraradices, a supposed ancient asexual arbuscular mycorrhizal fungus. BMC Evol Biol. 2009;9(1):1-11.

[56] Dickson S, Smith SE. Cross walls in arbuscular trunk hyphae form after loss of metabolic activity. New Phytol. 2001;151(3):735-42.

[57] Genre A, Chabaud M, Timmers T, Bonfante P, Barker DG. Arbuscular mycorrhizal fungi elicit a novel intracellular apparatus in *Medicago truncatula* root epidermal cells before infection. Plant Cell. 2005;17(12): 3489-99.

[58] Sanders IR. Specificity in the Arbuscular Mycorrhizal Symbiosis. 2002;157:415-37.

[59] Jacquemyn H, Merckx V, Brys R, Tyteca D, Cammue BPA, Honnay O, et al. Analysis of network architecture reveals phylogenetic constraints on mycorrhizal specificity in the genus Orchis (Orchidaceae). New Phytol. 2011;192(2):518-28.

[60] Hao Z, Xie W, Chen B. viruses Arbuscular Mycorrhizal Symbiosis A ff ects Plant. Viruses. 2019;11(534):1-12.

[61] Chifflot V, Rivest D, Olivier A, Cogliastro A, Khasa D. Molecular analysis of arbuscular mycorrhizal community structure and spores distribution in tree-based intercropping and forest systems. Agric Ecosyst Environ. 2009;131(1-2):32-9.

[62] Tawaraya K. Arbuscular mycorrhizal dependency of different plant species and cultivars. Soil Sci Plant Nutr. 2003;49(5):655-68.

[63] Bossou LR, Houngnandan HB, Adandonon A, Zoundji C. Diversité des champignons mycorhiziens arbusculaires associés à la culture du maïs (*Zea mays* L .) au Bénin Diversity of arbuscular mycorrhizal fungi associated with maize cropping (*Zea mays* L .) in Benin. 2019;13(April): 597-609.

[64] Sukmawati S, Adnyana A, Suprapta DN, Proborini M, Soni P, Adinurani PG. Multiplication arbuscular mycorrhizal fungi in Corn (*Zea mays* L.) with pots culture at greenhouse. E3S Web Conf. 2021; 226:1-10.

[65] Na Bhadalung N, Suwanarit A, Dell B, Nopamornbodi O, Thamchaipenet A, Rungchuang J. Effects of long-term NP-fertilization on abundance and diversity of arbuscular mycorrhizal fungi under a maize cropping system. Plant Soil. 2005;270(1):371-82.

[66] Oliveira CA, Sá NMH, Gomes EA, Marriel IE, Scotti MR, Guimarães CT, et al. Assessment of the mycorrhizal community in the rhizosphere of maize (*Zea mays* L.) genotypes contrasting for phosphorus efficiency in the acid savannas of Brazil using denaturing gradient gel electrophoresis (DGGE). Appl Soil Ecol [Internet]. 2009 Mar;41(3):249-58. Available from: https://linkinghub.elsevier.com/retrieve/pii/S0929139308001789

[67] Isobe K, Aizawa E, Iguchi Y, Ishii R. Distribution of arbuscular mycorrhizal fungi in upland field soil of Japan 1. Relationship between spore density and the soil environmental factor. Plant Prod Sci. 2007;10(1):122-8.

[68] Toljander JF, Santos-González JC, Tehler A, Finlay RD. Community analysis of arbuscular mycorrhizal fungi and bacteria in the maize mycorrhizosphere in a long-term fertilization trial. FEMS Microbiol Ecol. 2008;65(2):323-38.

[69] Sasvári Z, Hornok L, Posta K. The community structure of arbuscular mycorrhizal fungi in roots of maize grown in a 50-year monoculture. Biol Fertil Soils. 2011;47(2):167-76.

[70] An GH, Kobayashi S, Enoki H, Sonobe K, Muraki M, Karasawa T, et al. How does arbuscular mycorrhizal colonization vary with host plant genotype? An example based on maize (*Zea mays*) germplasms. Plant Soil. 2010;327(1):441-53.

[71] Hijri I, Sýkorová Z, Oehl F, Ineichen K, Mäder P, Wiemken A, et al. Communities of arbuscular mycorrhizal fungi in arable soils are not necessarily low in diversity. Mol Ecol. 2006;15(8): 2277-89.

[72] Alguacil M, Lumini E, Roldan A, Salinas-Garcia J, Bonfante P, Biaciotto V. the Impact of Tillage Practices on Arbuscular Mycorrhizal. Ecol Appl. 2008;18(2):527-36.

[73] Güimil S, Chang HS, Zhu T, Sesma A, Osbourn A, Roux C, et al. Comparative transcriptomics of rice reveals an ancient pattern of response to microbial colonization. Proc Natl Acad Sci U S A. 2005;102(22):8066-70.

[74] Breuillin F, Schramm J, Hajirezaei M, Ahkami A, Favre P,

Druege U, et al. Phosphate systemically inhibits development of arbuscular mycorrhiza in *Petunia hybrida* and represses genes involved in mycorrhizal functioning. Plant J. 2010;64(6): 1002-17.

[75] Xue L, Cui H, Buer B, Vijayakumar V, Delaux PM, Junkermann S, et al. Network of GRAS transcription factors involved in the control of arbuscule development in Lotus japonicus. Plant Physiol. 2015; 167(3):854-71.

[76] Gaude N, Bortfeld S, Duensing N, Lohse M, Krajinski F. Arbuscule-containing and non-colonized cortical cells of mycorrhizal roots undergo extensive and specific reprogramming during arbuscular mycorrhizal development. Plant J. 2012;69(3): 510-28.

[77] Ruzicka D, Chamala S, Barrios-Masias FH, Martin F, Smith S, Jackson LE, et al. Inside Arbuscular Mycorrhizal Roots – Molecular Probes to Understand the Symbiosis. Plant Genome. 2013;6(2):1-13.

[78] Willmann M, Gerlach N, Buer B, Polatajko A, Nagy R, Koebke E, et al. Mycorrhizal phosphate uptake pathway in maize: Vital for growth and cob development on nutrient poor agricultural and greenhouse soils. Front Plant Sci. 2013;4(DEC):1-6.

[79] Akiyama K, Ogasawara S, Ito S, Hayashi H. Structural requirements of strigolactones for hyphal branching in AM fungi. Plant Cell Physiol. 2010; 51(7):1104-17.

[80] Hogekamp C, Küster H. A roadmap of cell-type specific gene expression during sequential stages of the arbuscular mycorrhiza symbiosis. BMC Genomics. 2013;14(1).

[81] Blilou I, Bueno P, Ocampo JA, Garcia-Garrido JM. Induction of catalase and ascorbate peroxidase activities in tobacco roots inoculated with the arbuscular mycorrhizal Glomus mosseae. Mycol Res. 2000;104(6):722-5.

[82] Hogekamp C, Arndt D, Pereira PA, Becker JD, Hohnjec N, Küster H. Laser microdissection unravels cell-type-specific transcription in arbuscular mycorrhizal roots, including CAAT-Box transcription factor gene expression correlating with fungal contact and spread. Plant Physiol. 2011;157(4): 2023-43.

[83] Diédhiou I, Diouf D. Transcription factors network in root endosymbiosis establishment and development. World J Microbiol Biotechnol [Internet]. 2018;34(3):0. Available from: http://dx.doi.org/10.1007/s11274-018-2418-7

[84] Rich MK, Courty PE, Roux C, Reinhardt D. Role of the GRAS transcription factor ATA/RAM1 in the transcriptional reprogramming of arbuscular mycorrhiza in *Petunia hybrida*. BMC Genomics. 2017;18(1): 1-14.

[85] Liu F, Xu Y, Han G, Wang W, Li X, Cheng B. Identification and functional characterization of a maize phosphate transporter induced by mycorrhiza formation. Plant Cell Physiol. 2018;59(8):1683-94.

[86] Xu Y, Liu F, Li X, Cheng B. The mycorrhiza-induced maize ZmPt9 gene affects root development and phosphate availability in nonmycorrhizal plant. Plant Signal Behav [Internet]. 2018;13(12):1-3. Available from: https://doi.org/10.1080/15592324.2018.1542240

[87] Wang F, Cui PJ, Tian Y, Huang Y, Wang HF, Liu F, et al. Maize ZmPT7 regulates Pi uptake and redistribution which is modulated by phosphorylation. Plant Biotechnol J. 2020;18(12):2406-19.

[88] Rouf Shah T, Prasad K, Kumar P. Maize—A potential source of human nutrition and health: A review. Cogent Food Agric. 2016;2(1).

[89] Chávez-Arias CC, Ligarreto-Moreno GA, Ramírez-Godoy A, Restrepo-Díaz H. Maize Responses Challenged by Drought, Elevated Daytime Temperature and Arthropod Herbivory Stresses: A Physiological, Biochemical and Molecular View. Front Plant Sci. 2021;12(July):1-14.

[90] Kapoor R, Evelin H, Mathur P, Giri B. Arbuscular Mycorrhiza: Approaches for Abiotic Stress Tolerance in Crop Plants for Sustainable Agriculture. In: Plant Acclimation to Environmental Stress [Internet]. New York, NY: Springer New York; 2013. p. 359-401. Available from: http://link.springer.com/10.1007/978-1-4614-5001-6_14

[91] Pozo MJ, Jung SC, Martínez-Medina A, López-Ráez JA, Azcón-Aguilar C, Barea J-M. Root Allies: Arbuscular Mycorrhizal Fungi Help Plants to Cope with Biotic Stresses. In 2013. p. 289-307. Available from: http://link.springer.com/10.1007/978-3-642-39317-4

[92] Hoeksema JD, Chaudhary VB, Gehring CA, Johnson NC, Karst J, Koide RT, et al. A meta-analysis of context-dependency in plant response to inoculation with mycorrhizal fungi. Ecol Lett. 2010;13(3):394-407.

[93] Kaldorf M, Kuhn AJ, Schröder WH, Hildebrandt U, Bothe H. Selective element deposits in maize colonized by a heavy metal tolerance conferring arbuscular mycorrhizal fungus. J Plant Physiol. 1999;154(5-6):718-28.

[94] Estrada B, Aroca R, Maathuis FJM, Barea JM, Ruiz-Lozano JM. Arbuscular mycorrhizal fungi native from a Mediterranean saline area enhance maize tolerance to salinity through improved ion homeostasis. Plant, Cell Environ. 2013;36(10):1771-82.

[95] Lenoir I, Fontaine J, Lounès-Hadj Sahraoui A. Arbuscular mycorrhizal fungal responses to abiotic stresses: A review. Phytochemistry [Internet]. 2016;123:4-15. Available from: http://dx.doi.org/10.1016/j.phytochem.2016.01.002

[96] Gerlach N, Schmitz J, Polatajko A, Schlüter U, Fahnenstich H, Witt S, et al. An integrated functional approach to dissect systemic responses in maize to arbuscular mycorrhizal symbiosis. Plant, Cell Environ. 2015;38(8):1591-612.

[97] Patanita M, Campos MD, Félix MDR, Carvalho M, Brito I. Effect of tillage system and cover crop on maize mycorrhization and presence of Magnaporthiopsis maydis. Biology (Basel). 2020;9(3).

[98] Klaubauf S, Tharreau D, Fournier E, Groenewald JZ, Crous PW, de Vries RP, et al. Resolving the polyphyletic nature of Pyricularia (Pyriculariaceae). Stud Mycol [Internet]. 2014;79(1):85-120. Available from: http://dx.doi.org/10.1016/j.simyco.2014.09.004

[99] Molinero-Ruiz L, Melero-Vara J, Mateos A. Cephalosporium maydis, the Cause of Late Wilt in Maize, a Pathogen New to Portugal and Spain. Plant Dis - PLANT DIS. 2010 Mar 1;94:379.

[100] Drori R, Sharon A, Goldberg D, Rabinovitz O, Degani O, Mediterranea SP, et al. Molecular diagnosis for Harpophora maydis, the cause of maize late wilt in Israel Published by: Firenze University Press on behalf of the Mediterranean Phytopathological Union Stable URL: https://www.jstor.org/stable/42685381Moleculardiagnosis for Ha. 2013;52(1):16-29.

[101] Owolade OF, Alabi BS, Enikuomehin OA, Atungwu JJ. Effect of harvest stage and drying methods on germination and seed-borne fungi of maize (*Zea mays* L.) in South West Nigeria. African J Biotechnol. 2005;4(12):1384-9.

[102] Hussain N, Hussain A, Ishtiaq M, Azam S, Hussain T. Pathogenicity of two seed-borne fungi commonly involved in maize seeds of eight districts of Azad Jammu and Kashmir, Pakistan. African J Biotechnol. 2013;12(12): 1363-70.

[103] Olawuyi OJ, Odebode AC, Olakojo SA, Popoola OO, Akanmu AO, Izenegu JO. Host-pathogen interaction of maize (*Zea mays* L.) and Aspergillus niger as influenced by arbuscular mycorrhizal fungi (Glomus deserticola). Arch Agron Soil Sci [Internet]. 2014;60(11):1577-91. Available from: http://dx.doi.org/10.1080/03650340.2014.902533

[104] Olowe OM, Olawuyi OJ, Sobowale AA, Odebode AC. Role of arbuscular mycorrhizal fungi as biocontrol agents against Fusarium verticillioides causing ear rot of *Zea mays* L. (Maize). Curr Plant Biol [Internet]. 2018;15(November):30-7. Available from: https://doi.org/10.1016/j.cpb.2018.11.005

[105] Prasetyo J, Ginting C, Akin HM, Suharjo R, Niswati A, Afandi A, et al. The effect of biological agent and botanical fungicides on maize downy mildew. Biodiversitas. 2021;22(4): 1652-7.

[106] Atera EA, Itoh K, Azuma T, Ishii T. Farmers' perspectives on the biotic constraint of *Striga hermonthica* and its control in western Kenya. Weed Biol Manag. 2012;12(1): 53-62.

[107] Lendzemo VW, Van Ast A, Kuyper TW. Can arbuscular mycorrhizal fungi contribute to Striga management on cereals in Africa? Outlook Agric. 2006;35(4):307-11.

[108] Manjunatha PH, Nirmalnath PJ, Ht C. Field evalualtion of native arbuscular mycorrhizal fungi in the management of Striga in sugarcane (*Saccharum officinarum* L .). J Pharmacogn Phytochem. 2018;7(2): 2496-500.

[109] Othira, J. O. Effectiveness of arbuscular mycorrhizal fungi in protection of maize (*Zea mays* L.) against witchweed (*Striga hermonthica* Del Benth) infestation. J Agric Biotechnol Sustain Dev. 2012;4(3): 37-44.

[110] Bárzana G, Aroca R, Paz JA, Chaumont F, Martinez-Ballesta MC, Carvajal M, et al. Arbuscular mycorrhizal symbiosis increases relative apoplastic water flow in roots of the host plant under both well-watered and drought stress conditions. Ann Bot. 2012;109(5):1009-17.

[111] Mathur S, Jajoo A. Arbuscular mycorrhizal fungi protects maize plants from high temperature stress by regulating photosystem II heterogeneity. Ind Crops Prod [Internet]. 2020;143(November):111934. Available from: https://doi.org/10.1016/j.indcrop.2019.111934

[112] Wossen T, Abdoulaye T, Alene A, Feleke S, Menkir A, Manyong V. Measuring the impacts of adaptation strategies to drought stress: The case of drought tolerant maize varieties. J Environ Manage [Internet]. 2017;203:106-13. Available from: http://dx.doi.org/10.1016/j.jenvman.2017.06.058

[113] Mangena P. Water Stress: Morphological and Anatomical Changes in Soybean (*Glycine max* L.) Plants. Plant, Abiotic Stress Responses to Clim Chang. 2018;

[114] Brodribb TJ, Sussmilch F, McAdam SAM. From reproduction to production, stomata are the master regulators. Plant J. 2020;101(4): 756-67.

[115] Bahadur A, Batool A, Nasir F, Jiang S, Mingsen Q, Zhang Q, et al. Mechanistic insights into arbuscular mycorrhizal fungi-mediated drought stress tolerance in plants. Int J Mol Sci. 2019;20(17):1-18.

[116] Begum N, Ahanger MA, Su Y, Lei Y, Mustafa NSA, Ahmad P, et al. Improved Drought Tolerance by AMF Inoculation in Maize (*Zea mays*) Involves Physiological and Biochemical Implications. Plants [Internet]. 2019 Dec 6;8(12):579. Available from: https://www.mdpi.com/2223-7747/8/12/579

[117] Posta K, Hong Duc N. Benefits of Arbuscular Mycorrhizal Fungi Application to Crop Production under Water Scarcity. In: Drought - Detection and Solutions [Internet]. IntechOpen; 2020. p. 13. Available from: http://dx.doi.org/10.1039/C7RA00172J%0Ahttps://www.intechopen.com/books/advanced-biometric-technologies/liveness-detection-in-biometrics%0Ahttp://dx.doi.org/10.1016/j.colsurfa.2011.12.014

[118] Ruiz-Lozano JM, Aroca R, Zamarreño ÁM, Molina S, Andreo-Jiménez B, Porcel R, et al. Arbuscular mycorrhizal symbiosis induces strigolactone biosynthesis under drought and improves drought tolerance in lettuce and tomato. Plant Cell Environ. 2016;39(2):441-52.

[119] Saeed W, Naseem S, Ali Z. Strigolactones biosynthesis and their role in abiotic stress resilience in plants: A critical review. Front Plant Sci. 2017;8(August):1-13.

[120] Xu H, Shao H, Lu Y. Arbuscular mycorrhiza fungi and related soil microbial activity drive carbon mineralization in the maize rhizosphere. Ecotoxicol Environ Saf [Internet]. 2019;182(June):109476. Available from: https://doi.org/10.1016/j.ecoenv.2019.109476

[121] Subramanian KS, Vivek PN, Balakrishnan N, Nandakumar NB, Rajkishore SK. Effects of arbuscular mycorrhizal fungus Rhizoglomus intraradices on active and passive pools of carbon in long-term soil fertility gradients of maize based cropping system. Arch Agron Soil Sci [Internet]. 2019;65(4):549-65. Available from: https://doi.org/10.1080/03650340.2018.1512100

[122] Landberg R, Hanhineva K, Tuohy K, Garcia-Aloy M, Biskup I, Llorach R, et al. Biomarkers of cereal food intake. Genes Nutr. 2019;14(1):1-16.

[123] Chen A, Gu M, Wang S, Chen J, Xu G. Transport properties and regulatory roles of nitrogen in arbuscular mycorrhizal symbiosis. Semin Cell Dev Biol [Internet]. 2018;74:80-8. Available from: http://dx.doi.org/10.1016/j.semcdb.2017.06.015

[124] Santi C, Bogusz D, Franche C. Biological nitrogen fixation in non-legume plants. Ann Bot. 2013;111(5):743-67.

[125] Udvardi M, Poole PS. Transport and metabolism in legume-rhizobia symbioses. Annu Rev Plant Biol. 2013;64:781-805.

[126] Courty PE, Smith P, Koegel S, Redecker D, Wipf D. Inorganic Nitrogen Uptake and Transport in Beneficial Plant Root-Microbe Interactions. CRC Crit Rev Plant Sci. 2015;34(November 2014):4-16.

[127] van Velzen R, Holmer R, Bu F, Rutten L, van Zeijl A, Liu W, et al. Comparative genomics of the nonlegume Parasponia reveals insights into evolution of nitrogen-fixing rhizobium symbioses. Proc Natl Acad Sci U S A. 2018;115(20):E4700-9.

[128] Dellagi A, Quillere I, Hirel B. Beneficial soil-borne bacteria and fungi:

A promising way to improve plant nitrogen acquisition. J Exp Bot. 2020;71(15):4469-79.

[129] Ferlian O, Biere A, Bonfante P, Buscot F, Eisenhauer N, Fernandez I, et al. Growing Research Networks on Mycorrhizae for Mutual Benefits. Trends Plant Sci [Internet]. 2018;23(11):975-84. Available from: https://doi.org/10.1016/j.tplants.2018.08.008

[130] Jansa J, Forczek ST, Rozmoš M, Püschel D, Bukovská P, Hršelová H. Arbuscular mycorrhiza and soil organic nitrogen: network of players and interactions. Chem Biol Technol Agric [Internet]. 2019;6(1):1-10. Available from: https://doi.org/10.1186/s40538-019-0147-2

[131] Ferrol N, Azcón-Aguilar C, Pérez-Tienda J. Review: Arbuscular mycorrhizas as key players in sustainable plant phosphorus acquisition: An overview on the mechanisms involved. Plant Sci [Internet]. 2019;280(June):441-7. Available from: https://doi.org/10.1016/j.plantsci.2018.11.011

[132] Koegel S, Mieulet D, Baday S, Chatagnier O, Lehmann MF, Wiemken A, et al. Phylogenetic, structural, and functional characterization of AMT3;1, an ammonium transporter induced by mycorrhization among model grasses. Mycorrhiza. 2017;27(7):695-708.

[133] Giovannini L, Palla M, Agnolucci M, Avio L, Sbrana C, Turrini A, et al. Arbuscular mycorrhizal fungi and associated microbiota as plant biostimulants: Research strategies for the selection of the best performing inocula. Agronomy. 2020;10(1).

[134] Bertaux J, Schmid M, Prevost-Boure NC, Churin JL, Hartmann A, Garbaye J, et al. In situ identification of intracellular bacteria related to Paenibacillus spp. in the mycelium of the ectomycorrhizal fungus Laccaria bicolor S238N. Appl Environ Microbiol. 2003;69(7):4243-8.

[135] Budi SW, Van Tuinen D, Martinotti G, Gianinazzi S. Isolation from the *Sorghum bicolor* mycorrhizosphere of a bacterium compatible with arbuscular mycorrhiza development and antagonistic towards soilborne fungal pathogens. Appl Environ Microbiol. 1999;65(11):5148-50.

[136] Hildebrandt U, Ouziad F, Marner FJ, Bothe H. The bacterium *Paenibacillus validus* stimulates growth of the arbuscular mycorrhizal fungus Glomus intraradices up to the formation of fertile spores. FEMS Microbiol Lett. 2006;254(2):258-67.

[137] Wang C, White PJ, Li C. Colonization and community structure of arbuscular mycorrhizal fungi in maize roots at different depths in the soil profile respond differently to phosphorus inputs on a long-term experimental site. Mycorrhiza [Internet]. 2016; Available from: http://dx.doi.org/10.1007/s00572-016-0757-5

[138] Bukovská P, Bonkowski M, Konvalinková T, Beskid O, Hujslová M, Püschel D, et al. Utilization of organic nitrogen by arbuscular mycorrhizal fungi—is there a specific role for protists and ammonia oxidizers? Mycorrhiza. 2018;28(5-6):465.

[139] Gui H, Gao Y, Wang Z, Shi L, Yan K, Xu J. Arbuscular mycorrhizal fungi potentially regulate N2O emissions from agricultural soils via altered expression of denitrification genes. Sci Total Environ [Internet]. 2021;774:145133. Available from: https://doi.org/10.1016/j.scitotenv.2021.145133

[140] Storer K, Coggan A, Ineson P, Hodge A. Arbuscular mycorrhizal fungi reduce nitrous oxide emissions from

N2O hotspots. New Phytol. 2018;220(4):1285-95.

[141] Shen Y, Zhu B. Arbuscular mycorrhizal fungi reduce soil nitrous oxide emission. Geoderma [Internet]. 2021;402(April):115179. Available from: https://doi.org/10.1016/j.geoderma.2021.115179

[142] Bender SF, Conen F, Van der Heijden MGA. Mycorrhizal effects on nutrient cycling, nutrient leaching and N2O production in experimental grassland. Soil Biol Biochem [Internet]. 2015;80:283-92. Available from: http://dx.doi.org/10.1016/j.soilbio.2014.10.016

[143] Lazcano C, Barrios-Masias FH, Jackson LE. Arbuscular mycorrhizal effects on plant water relations and soil greenhouse gas emissions under changing moisture regimes. Soil Biol Biochem [Internet]. 2014;74:184-92. Available from: http://dx.doi.org/10.1016/j.soilbio.2014.03.010

[144] Bender SF, Plantenga F, Neftel A, Jocher M, Oberholzer HR, Köhl L, et al. Symbiotic relationships between soil fungi and plants reduce N2O emissions from soil. ISME J. 2014;8(6):1336-45.

[145] Zhang X, Wang L, Ma F, Shan D. Effects of Arbuscular Mycorrhizal Fungi on N2O Emissions from Rice Paddies. Water Air Soil Pollut. 2015;226(7):1-10.

[146] Cozzolino V, Di Meo V, Piccolo A. Impact of arbuscular mycorrhizal fungi applications on maize production and soil phosphorus availability. J Geochemical Explor [Internet]. 2013;129:40-4. Available from: http://dx.doi.org/10.1016/j.gexplo.2013.02.006

[147] Assogba SA, Adjovi NRA, Agbodjato NA, Sina H, Adjanohoun A, Baba-Moussa L. Evaluation of the Mixed Effects of Some Indigenous Strains of Arbuscular Mycorrhizal Fungi on the Growth of Maize Seedlings Under Greenhouse Conditions. Eur Sci J ESJ. 2020;16(3):275-94.

[148] Bárzana G, Aroca R, Bienert GP, Chaumont F, Ruiz-Lozano JM. New insights into the regulation of aquaporins by the arbuscular mycorrhizal symbiosis in maize plants under drought stress and possible implications for plant performance. Mol Plant-Microbe Interact. 2014;27(4):349-63.

[149] Quiroga G, Erice G, Aroca R, Chaumont F, Ruiz-Lozano JM. Enhanced drought stress tolerance by the arbuscular mycorrhizal symbiosis in a drought-sensitive maize cultivar is related to a broader and differential regulation of host plant aquaporins than in a drought-tolerant cultivar. Front Plant Sci. 2017;8(June):1-15.

[150] Ghorchiani M, Etesami H, Alikhani HA. Improvement of growth and yield of maize under water stress by co-inoculating an arbuscular mycorrhizal fungus and a plant growth promoting rhizobacterium together with phosphate fertilizers. Agric Ecosyst Environ [Internet]. 2018;258(February):59-70. Available from: https://doi.org/10.1016/j.agee.2018.02.016

[151] Celebi SZ, Demir S, Celebi R, Durak ED, Yilmaz IH. The effect of Arbuscular Mycorrhizal Fungi (AMF) applications on the silage maize (*Zea mays* L.) yield in different irrigation regimes. Eur J Soil Biol [Internet]. 2010;46(5):302-5. Available from: http://dx.doi.org/10.1016/j.ejsobi.2010.06.002

[152] Zhang W, Cao J, Zhang S, Wang C. Effect of earthworms and arbuscular mycorrhizal fungi on the microbial community and maize growth under salt stress. Appl Soil Ecol [Internet]. 2016;107:214-23. Available from: http://dx.doi.org/10.1016/j.apsoil.2016.06.005

[153] Xu H, Lu Y, Tong S. Effects of arbuscular mycorrhizal fungi on photosynthesis and chlorophyll fluorescence of maize seedlings under salt stress. Emirates J Food Agric. 2018;30(3):199-204.

[154] Hao L, Zhang Z, Hao B, Diao F, Zhang J, Bao Z, et al. Arbuscular mycorrhizal fungi alter microbiome structure of rhizosphere soil to enhance maize tolerance to La. Ecotoxicol Environ Saf [Internet]. 2021;212:111996. Available from: https://doi.org/10.1016/j.ecoenv.2021.111996

[155] He T, Li C. Harness the power of genomic selection and the potential of germplasm in crop breeding for global food security in the era with rapid climate change. Crop J [Internet]. 2020;8(5):688-700. Available from: https://doi.org/10.1016/j.cj.2020.04.005

[156] Kinghorn BP, Cowling WA, Li L, Siddique KHM, Banks RG. Modeling crop breeding for global food security during climate change. 2019;(July 2018):1-10.

[157] Yadav SS, Redden RJ, Hatfield JL, Ebert AW, Hunter D, Ortiz R. Role of Plant Breeding to Sustain Food Security under Climate Change. Food Secur Clim Chang. 2018;(December):145-58.

[158] Ramírez-flores MR, Perez-limón S, Li M, Barrales-gamez B. The genetic architecture of host response suggests a trade-off between mycorrhizal and non-mycorrhizal performance in field-grown maize. 2020;

[159] Fasusi OA, Amoo AE, Babalola OO. Propagation and characterization of viable arbuscular mycorrhizal fungal spores within maize plant (*Zea mays* L.). J Sci Food Agric. 2021;(March).

[160] Ren AT, Mickan BS, Li JY, Zhou R, Zhang XC, Ma MS, et al. Soil labile organic carbon sequestration is tightly correlated with the abundance and diversity of arbuscular mycorrhizal fungi in semiarid maize fields. L Degrad Dev. 2021;32(3):1224-36.

[161] Al-Maliki S, Al-Amery A, Sallal M, Radhi A, Al-Taey DKA. Effects of arbuscular mycorrhiza and organic wastes on soil carbon mineralisation, actinomycete sand nutrient content in maize plants (*Zea mays* l.). Malaysian J Soil Sci. 2021;25(December 2020):107-24.

[162] Vosátka M, Albrechtová J. Benefits of Arbuscular Mycorrhizal Fungi to Sustainable Crop Production. In: Microbial Strategies for Crop Improvement [Internet]. Berlin, Heidelberg: Springer Berlin Heidelberg; 2009. p. 205-25. Available from: http://link.springer.com/10.1007/978-3-642-01979-1_10

[163] Jacott CN. Trade-Offs in Arbuscular Mycorrhizal Symbiosis: Disease Resistance, Growth Responses and Perspectives for Crop Breeding. 2017;1-18.

[164] Li X, Quan X, Mang M, Neumann G, Melchinger A, Ludewig U. Flint maize root mycorrhization and organic acid exudates under phosphorus deficiency: Trends in breeding lines and doubled haploid lines from landraces. J Plant Nutr Soil Sci. 2021;184(3):346-59.

[165] Wang XX, van der Werf W, Yu Y, Hoffland E, Feng G, Kuyper TW. Field performance of different maize varieties in growth cores at natural and reduced mycorrhizal colonization: yield gains and possible fertilizer savings in relation to phosphorus application. Plant Soil. 2020;450(1-2):613-24.

[166] Chu Q, Wang X, Yang Y, Chen F, Zhang F, Feng G. Mycorrhizal responsiveness of maize (*Zea mays* L.) genotypes as related to releasing date and available P content in soil. Mycorrhiza. 2013;23(6):497-505.

[167] Eduardo Contreras-Liza S. Plant Breeding and Microbiome. Plant Breed - Curr Futur Views. 2021;(May).

[168] Ramírez-Flores MR, Perez-Limon S, Li M, Barrales-Gamez B, Albinsky D, Paszkowski U, et al. The genetic architecture of host response reveals the importance of arbuscular mycorrhizae to maize cultivation. Elife [Internet]. 2020 Nov 19;9:1-18. Available from: https://elifesciences.org/articles/61701

[169] Galluzzi G, Seyoum A, Halewood M, Noriega IL, Welch EW. The role of genetic resources in breeding for climate change: The case of public breeding programmes in eighteen developing countries. Plants. 2020; 9(9):1-19.

[170] Aguilar R, Carreón-Abud Y, López-Carmona D, Larsen J. Organic fertilizers alter the composition of pathogens and arbuscular mycorrhizal fungi in maize roots. J Phytopathol. 2017;165(7-8):448-54.

[171] Kaeppler SM, Parke JL, Mueller SM, Senior L, Stuber C, Tracy WF. Variation among maize inbred lines and detection of quantitative trait loci for growth at low phosphorus and responsiveness to arbuscular mycorrhizal fungi. Crop Sci. 2000;40(2):358-64.

[172] Ortiz R. Role of Plant Breeding to Sustain Food Security under Climate Change. Food Secur Clim Chang. 2018;(December):145-58.

[173] Xu Y, Liu X, Fu J, Wang H, Wang J, Huang C, et al. Enhancing Genetic Gain through Genomic Selection: From Livestock to Plants. Plant Commun [Internet]. 2020;1(1):100005. Available from: https://doi.org/10.1016/j.xplc.2019.100005

[174] Mores A, Borrelli GM, Laid G, Petruzzino G, Pecchioni N, Giuseppe L, et al. Genomic Approaches to Identify Molecular Bases of Crop Resistance to Diseases and to Develop Future Breeding Strategies. 2021;

[175] Singh RK, Prasad A, Muthamilarasan M, Parida SK, Prasad M. Breeding and biotechnological interventions for trait improvement: status and prospects. Planta [Internet]. 2020;252(4):1-18. Available from: https://doi.org/10.1007/s00425-020-03465-4

[176] Ahmar S, Gill RA, Jung KH, Faheem A, Qasim MU, Mubeen M, et al. Conventional and molecular techniques from simple breeding to speed breeding in crop plants: Recent advances and future outlook. Int J Mol Sci. 2020;21(7):1-24.

[177] Berger F, Gutjahr C. Factors affecting plant responsiveness to arbuscular mycorrhiza. Curr Opin Plant Biol [Internet]. 2021 Feb;59:101994. Available from: https://linkinghub.elsevier.com/retrieve/pii/S1369526620301527

[178] Jiang G-L. Molecular Markers and Marker-Assisted Breeding in Plants. In: Plant Breeding from Laboratories to Fields [Internet]. InTech; 2013. p. 45-83. Available from: http://www.intechopen.com/books/plant-breeding-from-laboratories-to-fields/molecular-markers-and-marker-assisted-breeding-in-plants

[179] Collard BCY, Mackill DJ. Marker-assisted selection: an approach for precision plant breeding in the twenty-first century. Philos Trans R Soc Lond B Biol Sci. 2008;363(1491):557-72.

[180] Badji A, Otim M, Machida L, Odong T, Kwemoi DB, Okii D, et al. Maize Combined Insect Resistance Genomic Regions and Their Co-localization With Cell Wall Constituents Revealed by Tissue-Specific QTL Meta-Analyses. Front

Plant Sci [Internet]. 2018;9(July). Available from: https://www.frontiersin.org/article/10.3389/fpls.2018.00895/full

[181] Sendek A, Karakoç C, Wagg C, Domínguez-Begines J, do Couto GM, van der Heijden MGA, et al. Drought modulatcs interactions between arbuscular mycorrhizal fungal diversity and barley genotype diversity. Sci Rep. 2019;9(1):1-15.

[182] Chu Q, Zhang L, Zhou J, Yuan L. Soil plant-available phosphorus levels and maize genotypes determine the phosphorus acquisition efficiency and contribution of mycorrhizal pathway. 2020;

[183] Nouri E, Surve R, Bapaume L, Stumpe M, Chen M, Zhang Y, et al. Phosphate Suppression of Arbuscular Mycorrhizal Symbiosis Involves Gibberellic Acid Signaling. Plant Cell Physiol. 2021;1-34.

[184] Allier A, Teyssèdre S, Lehermeier C, Moreau L, Charcosset A. Optimized breeding strategies to harness genetic resources with different performance levels. BMC Genomics. 2020;21(1):1DUMM.

[185] Breseghello F, Coelho ASG. Traditional and modern plant breeding methods with examples in rice (*Oryza sativa* L.). J Agric Food Chem. 2013;61(35):8277-86.

[186] Govindaraj M, Vetriventhan M, Srinivasan M. Importance of genetic diversity assessment in crop plants and its recent advances: An overview of its analytical perspectives. Genet Res Int. 2015;2015(Figure 1).

[187] Pradhan SK, Barik SR, Sahoo A, Mohapatra S, Nayak DK, Mahender A, et al. Population structure, genetic diversity and molecular marker-trait association analysis for high temperature stress tolerance in rice. PLoS One. 2016;11(8):1-23.

[188] Varshney RK, Bohra A, Yu J, Graner A, Zhang Q, Sorrells ME. Designing Future Crops: Genomics-Assisted Breeding Comes of Age. Trends Plant Sci [Internet]. 2021;26(6):631-49. Available from: https://doi.org/10.1016/j.tplants.2021.03.010

[189] Nadeem MA, Nawaz MA, Shahid MQ, Doğan Y, Comertpay G, Yıldız M, et al. DNA molecular markers in plant breeding: current status and recent advancements in genomic selection and genome editing. Biotechnol Biotechnol Equip [Internet]. 2017;2818:1-25. Available from: https://www.tandfonline.com/doi/full/10.1080/13102818.2017.1400401

[190] Collard BCY, Mackill DJ. Marker-assisted selection : an approach for precision plant breeding in the twenty-first century Marker-assisted selection: an approach for precision plant breeding in the twenty-first century. Phil Trans R Soc B. 2008;363(8):557-72.

[191] Wang X, Xu Y, Hu Z, Xu C. Genomic selection methods for crop improvement: Current status and prospects. Crop J [Internet]. 2018;6(4):1-11. Available from: https://doi.org/10.1016/j.cj.2018.03.001

[192] Jannink J, Lorenz AJ, Iwata H. Genomic selection in plant breeding: from theory to practice. Brief Funct Genomics. 2010;9(2):166-77.

[193] Nakaya A, Isobe SN. Will genomic selection be a practical method for plant breeding ? Ann Bot. 2012;1303-16.

[194] Lang M, Li X, Zheng C, Li H, Zhang J. Shading mediates the response of mycorrhizal maize (*Zea mays* L.) seedlings under varying levels of phosphorus. Appl Soil Ecol [Internet]. 2021;166(October):104060. Available from: https://doi.org/10.1016/j.apsoil.2021.104060

[195] Liu S, Guo X, Feng G, Maimaitiaili B, Fan J, He X. Indigenous

arbuscular mycorrhizal fungi can alleviate salt stress and promote growth of cotton and maize in saline fields. Plant Soil. 2016;398(1-2):195-206.

[196] Wang H, Liang L, Liu B, Huang D, Liu S, Liu R, et al. Arbuscular mycorrhizas regulate photosynthetic capacity and antioxidant defense systems to mediate salt tolerance in maize. Plants. 2020;9(11):1-17.

[197] Barber NA, Kiers ET, Theis N, Hazzard R V., Adler LS. Linking agricultural practices, mycorrhizal fungi, and traits mediating plant-insect interactions. Ecol Appl. 2013;23(7): 1519-30.

[198] Araus JL, Kefauver SC, Zaman-Allah M, Olsen MS, Cairns JE. Translating High-Throughput Phenotyping into Genetic Gain. Trends Plant Sci. 2018;23(5):451-66.

[199] Chawade A, Van Ham J, Blomquist H, Bagge O, Alexandersson E, Ortiz R. High-throughput field-phenotyping tools for plant breeding and precision agriculture. Agronomy. 2019;9(5):1-18.

[200] Li D, Quan C, Song Z, Li X, Yu G, Li C, et al. High-Throughput Plant Phenotyping Platform (HT3P) as a Novel Tool for Estimating Agronomic Traits From the Lab to the Field. Front Bioeng Biotechnol. 2021;8(January): 1-24.

[201] Watts-Williams SJ, Jewell N, Brien C, Berger B, Garnett T, Cavagnaro TR. Using high-throughput phenotyping to explore growth responses to mycorrhizal fungi and zinc in three plant species. Plant Phenomics. 2019;2019.

[202] Tran BTT, Cavagnaro TR, Jewell N, Brien C, Berger B, Watts-Williams SJ. High-throughput phenotyping reveals growth of *Medicago truncatula* is positively affected by arbuscular mycorrhizal fungi even at high soil phosphorus availability. Plants, People, Planet. 2021;3(5):600-13.

[203] Berger B, de Regt B, Tester M. Applications of High-Throughput Plant Phenotyping to Study Nutrient Use Efficiency. In: Plant Mineral Nutrients [Internet]. 2013. p. 277-90. Available from: internal-pdf://217.22.206.224/Alexou-2013-Methods for xylem sap collection3.pdf internal-pdf://003676 0810/Alexou-2013-Methods for xylem sap collection.pdf internal-pdf://1843399444/Alexou-2013-Methods for xylem sap collection1.pdf internal-pdf://164926113

[204] Xu Y, Li P, Yang Z, Xu C. Genetic mapping of quantitative trait loci in crops. Crop J [Internet]. 2017;5(2):175-84. Available from: http://dx.doi.org/10.1016/j.cj.2016.06.003

[205] Kibe M, Nyaga C, Nair SK, Beyene Y, Das B, Suresh LM, et al. Combination of Linkage Mapping, GWAS , and GP to Dissect the Genetic Basis of Common Rust Resistance in Tropical Maize Germplasm. Int J Mol Sci. 2020;21(6518).

[206] Arrones A, Vilanova S, Plazas M, Mangino G, Pascual L, Díez MJ, et al. The dawn of the age of multi-parent magic populations in plant breeding: Novel powerful next-generation resources for genetic analysis and selection of recombinant elite material. Biology (Basel). 2020;9(8):1-25.

[207] Gage JL, Monier B, Giri A, Buckler ES. Ten years of the maize nested association mapping population: Impact, limitations, and future directions. Plant Cell. 2020;32(7): 2083-93.

[208] Korte A, Farlow A. The advantages and limitations of trait analysis with GWAS: a review. Plant Methods [Internet]. 2013;9(1):1-9. Available from: Plant Methods

[209] Stearns FW. One hundred years of pleiotropy: A retrospective. Genetics. 2010;186(3):767-73.

[210] Jia Y, Jannink JL. Multiple-trait genomic selection methods increase genetic value prediction accuracy. Genetics. 2012;192(4):1513-22.

[211] Maier RM, Zhu Z, Lee SH, Trzaskowski M, Ruderfer DM, Stahl EA, et al. Improving genetic prediction by leveraging genetic correlations among human diseases and traits. Nat Commun [Internet]. 2018;9(1):1-17. Available from: http://dx.doi.org/10.1038/s41467-017-02769-6

[212] Yu J, Pressoir G, Briggs WH, Bi IV, Yamasaki M, Doebley JF, et al. A unified mixed-model method for association mapping that accounts for multiple levels of relatedness. Nat Genet. 2006;38(2):203-8.

[213] Jaiswal V, Gahlaut V, Meher PK, Mir RR, Jaiswal JP, Rao AR, et al. Genome wide single locus single trait, multi-locus and multi-trait association mapping for some important agronomic traits in common wheat (*T. aestivum* L.). PLoS One. 2016;11(7):e0159343.

[214] Gupta PK, Kulwal PL, Jaiswal V. Association mapping in plants in the post-GWAS genomics era. In: Advances in Genetics [Internet]. 1st ed. Elsevier Inc.; 2019. p. 1-80. Available from: http://dx.doi.org/10.1016/bs.adgen.2018.12.001

[215] Zhang Y, Jia Z, Dunwell JM. Editorial: The Applications of New Multi-Locus GWAS Methodologies in the Genetic Dissection of Complex Traits. 2019;10(February):1-6.

[216] Fernandes SB, Zhang KS, Jamann TM, Lipka AE. How Well Can Multivariate and Univariate GWAS Distinguish Between True and Spurious Pleiotropy? Front Genet. 2021; 11(January):1-11.

[217] Momen M, Mehrgardi AA, Roudbar MA, Kranis A, Pinto RM, Valente BD, et al. Including phenotypic causal networks in genome-wide association studies using mixed effects structural equation models. bioRxiv [Internet]. 2018 Oct 9;9(October): 251421. Available from: https://www.frontiersin.org/article/10.3389/fgene.2018.00455/full

[218] Verhulst B, Maes HH, Neale MC. GW - SEM: A Statistical Package to Conduct Genome - Wide Structural Equation Modeling. Behav Genet. 2017;0(0):0.

[219] Momen M, Campbell MT, Walia H, Morota G. Utilizing trait networks and structural equation models as tools to interpret multi-trait genome-wide association studies. Plant Methods [Internet]. 2019;15(1):1-14. Available from: https://doi.org/10.1186/s13007-019-0493-x

[220] Grotzinger AD, Rhemtulla M, de Vlaming R, Ritchie SJ, Mallard TT, Hill WD, et al. Genomic SEM Provides Insights into the Multivariate Genetic Architecture of Complex Traits. bioRxiv [Internet]. 2018; Available from: http://biorxiv.org/content/early/2018/04/21/305029.abstract

[221] Igolkina AA, Meshcheryakov G, Gretsova M V., Nuzhdin S V., Samsonova MG. Multi-trait multi-locus SEM model discriminates SNPs of different effects. BMC Genomics [Internet]. 2020;21(Suppl 8):1-11. Available from: http://dx.doi.org/10.1186/s12864-020-06833-2

[222] Badji A, Kwemoi DB, Machida L, Okii D, Mwila N, Agbahoungba S, et al. Genetic Basis of Maize Resistance to Multiple Insect Pests: Integrated Genome-Wide Comparative Mapping and Candidate Gene Prioritization. Genes (Basel) [Internet]. 2020 Jun 24;11(6):689. Available from: https://www.mdpi.com/2073-4425/11/6/689

[223] Hassani-Pak K, Rawlings C. Knowledge Discovery in Biological Databases for Revealing Candidate Genes Linked to Complex Phenotypes. J Integr Bioinform. 2017;14(1):1-9.

[224] Muthuramalingam P, Krishnan SR, Pothiraj R. Global Transcriptome Analysis of Combined Abiotic Stress Signaling Genes Unravels Key Players in *Oryza sativa* L.: An In silico Approach. 2017;8(May):1-13.

[225] Woldesemayat AA, Modise DM, Gemeildien J, Ndimba BK, Christoffels A. Cross-species multiple environmental stress responses: An integrated approach to identify candidate genes for multiple stress tolerance in sorghum (*Sorghum bicolor* (L.) Moench) and related model species. PLoS One. 2018;13(3):1-30.

[226] Qaim M. Role of New Plant Breeding Technologies for Food Security and Sustainable Agricultural Development. Appl Econ Perspect Policy. 2020;42(2):129-50.

[227] Andersen EJ, Ali S, Byamukama E, Yen Y. Disease Resistance Mechanisms in Plants. 2018;

[228] Erb M, Reymond P. Molecular Interactions Between Plants and Insect Herbivores. Annu Rev Plant Biol. 2019;70(1):527-57.

[229] Hickey JM, Chiurugwi T, Mackay I, Powell W. Genomic prediction unifies animal and plant breeding programs to form platforms for biological discovery. Nat Genet. 2017;49(9):1297-303.

[230] Azodi CB, McCarren A, Roantree M, Campos G de los, Shiu S-H. Benchmarking algorithms for genomic prediction of complex traits. bioRxiv [Internet]. 2019;614479. Available from: https://www.biorxiv.org/content/10.1101/614479v1.full

[231] Moeinizade S, Hu G, Wang L, Schnable PS. Optimizing Selection and Mating in Genomic Selection with a Look-Ahead Approach: An Operations Research Framework. G3: Genes|Genomes|Genetics. 2019;9(7):2123-33.

[232] Montesinos-López OA, Montesinos-López A, Crossa J, Toledo FH, Pérez-Hernández O, Eskridge KM, et al. A Genomic Bayesian Multi-trait and Multi-environment Model. G3: Genes|Genomes|Genetics [Internet]. 2016 Sep;6(9):2725-44. Available from: http://g3journal.org/lookup/doi/10.1534/g3.116.032359

[233] Gill HS, Halder J, Zhang J, Brar NK, Rai TS, Hall C, et al. Multi-Trait Multi-Environment Genomic Prediction of Agronomic Traits in Advanced Breeding Lines of Winter Wheat. Front Plant Sci. 2021;12(August):1-14.

[234] Montesinos-López OA, Montesinos-López A, Tuberosa R, Maccaferri M, Sciara G, Ammar K, et al. Multi-Trait, Multi-Environment Genomic Prediction of Durum Wheat With Genomic Best Linear Unbiased Predictor and Deep Learning Methods. Front Plant Sci [Internet]. 2019;10(November):1311. Available from: https://www.frontiersin.org/article/10.3389/fpls.2019.01311/full

[235] Montesinos-López OA, Montesinos-López A, Crossa J, Gianola D, Hernández-Suárez CM, Martín-Vallejo J. Multi-trait, multi-environment deep learning modeling for genomic-enabled prediction of plant traits. G3 Genes, Genomes, Genet. 2018;8(12):3829-40.

[236] de Oliveira AA, Resende MFR, Ferrão LFV, Amadeu RR, Guimarães LJM, Guimarães CT, et al. Genomic prediction applied to multiple traits and environments in second season maize hybrids. Heredity (Edinb) [Internet]. 2020;125(1-2):60-72. Available from: http://dx.doi.org/10.1038/s41437-020-0321-0

Chapter 5

Aflatoxins and Fumonisins Contamination of Maize in Bangladesh: An Emerging Threat for Safe Food and Food Security

Muhtarima Jannat, Md. Mostafa Masud, Mushfika Nusrat, Samrin Bashar, Mamuna Mahjabin Mita, Muhammad Iqbal Hossain, Md. Zahangir Alam, Sabina Yeasmin and Md. Rashidul Islam

Abstract

Maize (Bhutta) is one of the important growing cereal crops in Bangladesh. Toxigenic fungi such as *Aspergillus* and *Fusarium* infect stored maize grains. Enzyme-linked immusorbent assay (ELISA) was used to determine total aflatoxins and fumonisins contamination in stored maize grains collected from 15 Bangladeshi maize-producing areas. The highest total concentration of aflatoxins (103.07 μg/kg) and fumonisin (9.18 mg/kg) was found in Chuadanga and Gaibandha, whereas the lowest was detected for aflatoxins (1.07 μg/kg) and (0.11 mg/kg) in Dinajpur and Cumilla, respectively. The findings clearly demonstrated that aflatoxin concentrations in samples from six regions and fumonisin concentrations in samples from 10 regions were beyond the regulatory limit of aflatoxin (10 ppb) and fumonisin (1 ppm), respectively, as set by European Union (EU). However, a positive correlation between aflatoxins with toxigenic *A. flavus*, and fumonisins with toxigenic *Fusarium* spp. was observed. The fungi associated with maize grains were identified by sequencing of ITS regions. Moreover, toxigenic *A. flavus* was confirmed using primers specific to *nor*, *apa2*, *omtA* and primer FUM1 for *F. proliferatum* and *F. oxysporum*. Since the Bangladesh Food Safety Authority has not authorized any precise regulation limits for maize mycotoxin contamination, these results will serve as a benchmark for monitoring mycotoxin contamination in maize and also to develop globally practiced biocontrol approach for producing safe food and feed.

Keywords: mycotoxins, maize, threat, food, security

1. Introduction

Maize (Bhutta) or *Zea mays L.* (corn) is one of the supreme vital cereals in the globe which belongs to Poaceae family and it has been ranked as a third position in the last few decades after wheat and rice [1]. A fair number of food and industrial commodities such as maize flour, animal feed, cooking ingredient, corn syrup, grain alcohol and whiskey are processed from maize [2]. Maize has been known as a significant emerging crop in Bangladesh as well as maize production is familiarized day by day due to its diverse use for feed, food, fish meal and edible oil processing [3]. Bangladesh has achieved 11th position when it comes to average yield which was 8 tons per ha in the year of 2019–2020 [4] and maize production were 40 lakh ton [5]. Anyway, maize plant is quite vulnerable for various fungi as they get favorable environment to infect via fluctuation of humidity and temperature conditions in both of storage and growing phase [6]. In harvesting period less care in drying and storage processing leads to a surge in infection and production of toxin [7]. Dominant pathogens such as *Aspergillus* spp. and *Fusarium* spp. in maize have the capability to destroy seeds, germination procedure in seeds as well as generating vital mycotoxins [8]. Mycotoxins are light molecular weight developed from saprophytic fungi, most significantly *Aspergillus*, *Fusarium* and *Penicillium* as secondary metabolites [9]. Mycotoxins were detected as one of the deadly toxins after the outbreak of ruinous 'Turkey X' in 1960s at England which leads to the death of Turkey poults (100,000) [10]. Mycotoxin comtamination can develop in any stage of food chain especially in the field, during transportation, processing, harvesting and storage [11].

Aflatoxins are mainly hepatocarcinogenic toxins comprising of major three metabolities named Aflatoxin G, M and B under derivative compounds named difurocoumarin [12–14]. The paramount aflatoxin producing fungi globally is *A. flavus* divided into two distinct morphotypes named L and S [15], among them S morphotype was potentially ruinous as it was capable of producing gigantic level of toxins [16, 17]. A significant research has been made by toxigenic communities that innumerable lineages of fungi are belong to S morphotype among them a few were able to engender enormous concentration of both B and G aflatoxins [18]. Several *Aspergillus* spp. is accounted for several toxins such as aflatoxin B is mainly produced from *A. flavus, A. parasiticus* whereas aflatoxins G is developed from *A. nomius.* Moreover, G and B are highly produced inspices, fruits, corn, nuts, peanuts and copra [19, 20]. *A. flavus* is ubiquitous and mostly detected in corn producing toxins, while in peanut *A. parasiticus* is the main culprit of developing toxins [21]. The toxicity level of aflatoxins of different types chronologically are B1 > G1 > B2 > G2 [22]. Basically, aflatoxins levels were found ascendency in the food markets of Bangladesh [23]. Temperature, pH, relative humidity, and the presence of other fungi are predominant factor for developing aflatoxins and substrates [24]. Aflatoxins level surges due to drought, insect damage, and heat during fungal growth [25]. The *AflR* gene regulates the activation of other structural genes including *omt-A*, *ver-1*, and *nor-1*, which are involved in the aflatoxin biosynthesis process [26]. In hot and humid settings, aflatoxins contamination are also thrived [27]. Seasonal variation has been observed in Bangladesh including high humidity, high temperature and seasonal variation in rainfall (http://en.wikipedia.org/wiki/Geography of Bangladesh). Extreme humid conditions significantly triggered the growth of aflatoxins [28], as a result, it is obvious that aflatoxins was reported in maize, cereals and groundnuts and other feed in Bangladesh and exceeding European Union (EU) permissible limit for aflatoxins [29].

Fusarium spp. are among the utmost crucial fungal pathogens of maize, where they cause severe abatement of yield and accumulation of a vast range of harmful mycotoxins in the grain [30]. *Fusarium* spp. also have the ability to infect crucial crops such as potato, wheat, barley, asparagus, mango, oats, rice and other feed and food crops [31]. High moisture conditions triggered the production of *Fusarium* toxins near or at harvesting stage in cereals [32, 33]. Fumonisins toxins can be developed from a numerous species such as *F. moniliforme*, *F. verticillioides*, *F. nygamai*, *F. proliferatum* [34] as well as *A. niger* [35]. Fumonisins comprise of four types of toxins which are A, B, C, and P, among them fumonisin B1 is the most exploited and ruinous one [34]. FB1, FB2, and FB3 were designated as utmost destructive and highly abundant fumonisin toxins where FB1 is the most ruinous due to its availability of high concentration on host ranging from 70 % to 80 % of all fumonisins [36–38]. Several biotic (temperature, water stress) and abiotic (osmotic stress, pH, and fungicides) factors are resposnsible for *Fusarium* growth and Fumonisin production [39, 40]. At maturity stage damage occurs by insects, during flowering wet warm weather, rain before harvest, humidity, and media composition for both the *Fusarium* spp., all the activites are related to fumonisins production [41, 42]. *FUM1* gene can also expressed by ecological conditions reported by [43, 44]. As *Fusarium* is widespread and ubiquitous in all cereal growing regions of the globe and corresponding mycotoxins are produced which has been influenced by storage methods and crop production [45]. In the midst of milling, storage, processing, cooking of food and feed, *Fusarium* are highly stable due to its structure and humans and animals are exhibited to them to a certain degree [46–48]. In Bangladesh, animal feed samples were detected and found fumonisin contamination mainly maize based feed contamination [49].

An investigation came out that in South Asia has been ranking as the utmost prevalent continent in case of exposing aflatoxins contamination (82 %) in the globe as well as 41 % maize samples were detected higher amount of aflatoxins contamination than the permissible limit of lenient EU criteria [49]. The very first outbreak of mycotoxin (Sterigmatocystin) was found in Bangladesh in rice straw [50], later in maize and poultry birds [51]. Liver cancer and hepatitis B infection promotes carcinogenic potency in specific individuals by aflatoxins [52, 53]. In Japan, in the year of 1991–2009, violation cases were detected exceeded 1500 in foods which were imported at a level of 10–4918 mg/kg [54]. 62 % children with the age of 3 are at a complete risk of infecting with aflatoxins as aflatoxins biomarkers was detected in plasma of their blood [55]. According to WFP (World Food Program), permissible limit of aflatoxins is 10 ppb (10 μg/kg) and for fumonisins it is 1 ppm (1 mg/kg) [56]. Fumonisins toxin may causes esophageal carcinoma in humans [57], as well as contaminated with folate uptake in cellular level [58] and surging the intensity of neural tube defect [59]. 52 % positive rate of fumonisins was found with an overall level of 936 mg/kg in Asia [60]. *Fusarium* mycotoxin can cause leukoencephalomalacia, porcine pulmonary edema and rat hepatocarcinoma in human and livestock as well [55, 61, 62] detected that in Dhaka, Bangladesh 62 % of 3 year old children had aflatoxin biomarkers in their blood plasma revealing chronic aflatoxin exposure as reported earlier that significant amount aflatoxins were found from corn selling in the Bangladeshi market. Probably 1311 cases of liver cancer was detected every year in Bangladesh [63]. In can be deduced from abovementioned fact that determining aflatoxins and fumonisins and all other mycotoxins in food and feed are the prime need for the country like Bangladesh as these mycotoxin substantially subverts our plants yield concurrently human and animal lives as well. Thus, more research needs to be conducted to elicit the specific mycotoxin hampering specific food, feed and plants, besides to find out the plausible management for controlling these mycotoxins. This study highly exhibited the aflatoxins and fumonisins toxin level in Bangladesh from maize samples of different regions as it has been concerned as one of the burning issues for ensuring safety food.

2. Materials and methods

2.1 Sample collection

Composite stored maize grain samples were collected from 15 maize growing areas of Bangladesh such as Bogura, Kushtia, Meherpur, Chuadanga, Kishoreganj, Manikganj, Cumilla, Rajshahi, Dinajpur, Rangpur, Natore, Thakurgaon,

Panchagarh, Nilphamary and Jashore. Maize samples were collected from stores of traders in local markets of different districts. Ten markets were sampled in each district having at least five traders in each market. At least two quarter of kilogram unique samples were coalesced from each trader for laboratory analysis. Samples were collected after thoroughly mixing maize in the bag to increase chances of getting the fungi. The samples were stored at temperatures below 4° C to await analysis.

3. Detection of aflatoxins and fumonisins by ELISA method

3.1 Procedure of sample preparation

A representative sample was taken and it was grounded with blender so that 75 % of that grounded portion can pass through a 20-mesh sieve, then thoroughly the sub-sample portion was mixed. 50 g of ground sample was weighed out into a clean conical flask that can be tightly sealed. 250 mL of methanol (70 % methanol diluted in water) extraction solution was added to the ground sample and the flask was sealed. Then the conical flask containing the sample was shaken for 3 min. The sample was allowed to settle down, then the top layer of extract was filtered through a Whatman #1 filter paper and the filtrate sample was collected. The prepared extract was diluted at 1:20 with distilled water. Sample was ready for testing without further preparation.

3.2 Assay protocol for aflatoxins

200 μL conjugate solution was pipetted into dilution wells. 100 μL of each standard or sample extract was added into the dilution wells. The mixture was mixed well and 100 μL of the mixture (conjugate and standard or samples) was transferred into antibody-coated wells. The plate was then incubated for 15 min with slow shaking and washed with distilled water for 5 times. The plate was then tap dried. 100 μL of substrate solution was pipetted into antibody coated wells. The plate was incubated with shaking for 5 min. 100 μL of stop solution was pipetted into antibody coated wells. The absorbance of each well was read at 450 nm with a differential filter at 630 nm. As the aflatoxin limit was (0–40) ppb but we found more than that which was diluted by dilution factor in three regions (Bogura, Nilphamari, Rangpur) by four times dilution.

3.3 Assay procedure for fumonisins detection

200 μL conjugate solution was pipetted into dilution wells with 100 μL of each standard and sample extract. The mixture was mixed well and 100 μL of the mixture (conjugate and standard or samples) was transferred into antibody-coated wells. The plate was then incubated for 15 min with slow shaking and then washed with distilled water for 5 times. The plate was then tap dried. 100 μL of substrate solution was pipetted into antibody coated wells. The plate was incubated with shaking for 5 min. 100 μL of stop solution was pipetted into antibody coated wells. The absorbance of each well was read at 450 nm with a differential filter at 630 nm.

3.4 Isolation, purification, identification and preservation of mycotoxigenic fungi

Isolation & purification of *Aspergillus* spp. and *Fusarium* spp. were collected from stored maize grain samples which was conducted by blotter method [64, 65]. In this method, 400 maize grains were tested for the identification of toxigenic *Aspergillus* spp. and *Fusarium* spp. for each sample collected from different locations and 40 plastic pestridishes were used for each sample. Then 10 maize grains were placed in the sterile plastic petridish containing three layers of wet blotter papers. The petridish was incubated at 25 ± 1° C under 12/12 h light and darkness cycle for 7 days. Each seed was observed under stereo microscope (Stemi 508, Germany) in order to record the presence of fungal colonies and temporary slides were prepared from the fungal colonies for morphological identification under compound microscope (Primo Star, Germany). Or one of the quarter kilo samples from each trader milled into fine floor using a Laboratory Milling machine. Ten grams of the ground sample was mixed with 100 ml sterile water to make a stock solution and serially diluted up to dilution 10^3. The suspension was plated in Potato Dextrose Agar Medium (PDA) [66, 67] by mixing 1 ml suspension in molten PDA cooled to 40° C. Isolation media was prepared by weighing 39 g of PDA into 1 L of water. The mixture was autoclaved for 15 min at 121° C and 15 PSI pressure. The media was allowed to cool to about 50° C and then amended with 25 ng/L of streptomycin and tetracycline [68, 69]. Petri dishes were labeled appropriately and a milliliter of the diluted sample was poured into a sterile petri dish aseptically and then 18 ml of PDA media at 40° C will was poured on the same plate and the mixture swirled gently to mix. The mixture was allowed to cool and solidify in the laminar flow hood and then sealed using parafilm for incubation. The plates were incubated at room temperature for 5–7 days.

4. Molecular based identification of fungi

4.1 DNA extraction

Before DNA extraction each purified *Aspergillus* spp. and *Fusarium* spp. was grown on PDA for 7–10 days at 28° C in an incubator. Then a 5 mm culture block was transferred on the conical flask containing PDA broth and the flasks were incubated at 28° C in an incubator for 7–10 days. Mycelium of each isolate was harvested and preserved at −80° C.

Genomic DNA was extracted from the fungal species isolated from maize grains following Wizard Genomic DNA extraction kit (Promega, USA) according to the manufacturer instructions from 100 mg fungal tissue ground with liquid nitrogen. Fungal tissue was processed by freezing with liquid nitrogen and grinding into a fine powder using a microcentrifuge tube pestle or a mortar and it was pestled. 0.04 g of this fungal tissues powder was added to a 1.5 ml microcentrifuge tube. 600 μl of Nuclei Lysis Solution was added and it was vortexed for 1–3 s to wet the tissue. The sample was incubated at 65° C for 15 min. 3 μl of RNase Solution was added to the cell lysate, and the sample was mixed by inverting the tube 2–5 times. The mixture was incubated at 37° C for 15 min. The sample was allowed to cool to room temperature for 5 min before proceeding. 200 μl of Protein Precipitation Solution was added, and it was vortexed vigorously at high speed for 20 s. The sample was centrifuged for 3 min at 13,000–16,000 × *g*. The precipitated proteins were formed into a tight pellet. The supernatant was carefully removed containing the DNA (leaving the protein pellet behind) and it was transferred to a clean 1.5 ml microcentrifuge tube containing 600 μl of room temperature isopropanol. The solution was gently mixed by inversion until thread-like strands of DNA form a visible mass. Then the sample was centrifuged at 13,000–16,000 × *g* for 1 min at room temperature. The supernatant carefully decanted. 600 μl of room temperature 70 % ethanol was added and was inverted gently into the tube several times to wash the DNA. It was centrifuged at 13,000–16,000 × *g* for 1 min at room temperature. The ethanol was aspirated carefully using either a drawn Pasteur pipette or a sequencing pipette tip. The DNA pellet was very loose at this point and care must be used to avoid aspirating the pellet into the pipette. The tube was inverted onto clean absorbent paper and the pellet was air-dried for 15 min. 100 µl of DNA Rehydration Solution was added and the DNA was rehydrated by incubating at 65° C for 1 h. Periodically the solution was mixed by gently tapping the tube. Alternatively, the DNA was rehydrated by incubating the solution overnight at room temperature or at 4° C. The DNA was stored at 2–8° C.

4.2 Primers, PCR conditions and sequencing of ITS region

The extracted DNA samples were amplified with PCR reaction for ITS regions. The forward primer: ITS1-5.8S (5′-GGAAGTAAAAGTCGTAACAAGG-3′) and the reverse primer rDNA-ITS4 (TCCTCCGCTTATTGATATGC) were used [70]. PCRs were performed in 25 µl reaction volume containing 12.5 µl master mix, 1 µl ITS1, 1 µl ITS4, 9.5 µl Nuclease free water and 1 µl templet DNA (100 ng/µl). PCR products were visualized in 2 % agarose gel, dyed with ethidium bromide and the photograph was taken using a Gel documentation system (Dynamica, GelView Master). The conditions for PCR reaction was: initial denaturation for 5 min at 95° C, followed by 34 cycles at 95° C for 30s, at 55° C for 1 min and at 72° C for 1 min and then final elongation at 72° C for 6 min. The amplified products were stored at −20° C. PCR products were sequenced using ITS1 primer via commercial outsourcing at Macrogen, Korea via Biotech concern. Finaly, Sequence data were imported by Chromas Software version 2. Sequence data were analyzed by BLAST program (Basic Local Alignment Search Tool) and GenBank (https://blast.ncbi.nlm.nih.gov/Blast.cgi).

5. PCR based detection of aflatoxin producing *Aspergillus* spp

5.1 PCR primers and amplification

Primers *nor-1* FP (5′-ACCGCTACGCCGGCACTC TCGGCAC-3′) and *nor-1* RP (5′-GTTGGCCGCCAG CTTCGACACTCCG-3′) were set to amplify an amplicon of 400 bp of norsolorinic acid reductase; *omtA* FP (5′-GGCCCGGTTCCTTG GCTCCTAAGC3′) and *omtA* RP (5′-CGCCCCAGTGAGACCCTTCC TCG-3′) to amplify a 1024 bp fragment of sterigmatocystin O-methyltransferase; and *aflR* FP (5′-TATCT CCCCCCGGGCATCTCCCGG-3′) and *aflR* RP (5′-CCGTCAGACAGCCACTGGACACGG-3′) to amplify a 1032 bp fragment of regulatory protein (*aflR*), involved in aflatoxin biosynthesis. The nucleotide sequence of all these genes from *A. parasiticus* are available at NCBI, GenBank at aceession numbers L27801 (*nor-1*), SRRC 2043 (*aflR*) and SRRC 143 (*omt-1*). PCR was performed in 15 µL of reaction volume containing 7.5 µl master mix, 1 µl forward primer, 1 µl of reverse primer and 4.5 µl nuclease free water and 1 µl of extracted DNA as template (with a total concentration of 100 ng of genomic DNA per reaction). PCR condition for *nor 1* primer initial denaturation for 5 min at 94° C, followed by 35 cycles at 94° C for 30 s, at 67° C for 30 s and at 72° C for 30 s and then final elongation at 72° C for 10 min [71]. PCR condition for *omtA* and *aflR* primer initial denaturation for 10 min at 95° C, followed by 30 cycles at 94° C for 1 min, at 65° C for 2 min and at 72° C for 2 min and then final elongation at 72° C for 5 min [71]. PCR products were separated by electrophoresis on a 1 % agarose gel with 0.5 % ethidium bromide in 1× TBE buffer and visualized under a Gel documentation system (Dynamica, GelView Master). 1 kb plus DNA Ladder (BioLabs, New England) was used as molecular size marker for the analysis of fragment size.

6. PCR based identification of mycotoxigenic *Fusarium* spp

6.1 Primers for PCR amplification

Primers specific for fumonisins producing *Fusarium* spp. (*FUM1* Forward-CCATCAC AGTGGGACACAGT, *FUM1* Reverse-CGTATC GTCAGCATGATGTAGC) were used previously [72]. PCR were performed in mixture 15 μl volume containing 1 μl of DNA sample, 7.5 μl of master mix, 1 μl *FUM1* forward primer, 1 μl *FUM1* reverse primer, 4.5 μl nuclease free water. PCR was performed using T100 Thermocycler (BioRad, Hercules, USA). The PCR condition for *FUM1* regions include 94° C for 4 min for initial denaturation, followed by 35 cycles of denaturation at 94° C for 1 min, primer annealing at 58° C for 1 min, primer extension at 72° C for 1 min. The final extension was set at 72° C for 10 min. 4 μl of the PCR product was electrophoresed on 1.5 % agarose gel, stained with ethidium bromide, illuminated and documented using Gel documentation system (Dynamica, GelView Master).

7. Statistical analyses

The collected data were analyzed statistically by using Minitab software version 17 (www.minitab.com). The mean of all the treatments were compared by critical difference value at 5 % level of significance.

8. Results

8.1 Determination of total Aflatoxins contamination in stored maize grain samples collected from some selected growing areas of Bangladesh

The study was performed at the Laboratory of Department of Plant Pathology, Bangladesh Agricultural University, Mymensingh. Composite stored maize grain samples were collected from 15 maize growing areas of Bangladesh including Panchagarh, Thakurgaon, Dinajpur, Nilphamari, Rangpur, Lalmonirhat, Gaibandha, Bogura, Natore, Kushtia, Jashore, Chuadanga, Kishoreganj, Manikganj and Cumilla.

In terms of total aflatoxins concentration in μg/kg, the highest and lowest amount of aflatoxins concentration was recorded in Chuadanga (101.57 μg/kg) and Dinajpur (1.08 μg/kg) which exposed no significant relationship to each other. The moderate amount of afalatoxin level was detected in Gaibandha (68.73 μg/kg), Kushtia (31.48 μg/kg), Kishoreganj (30.86 μg/kg), Rangpur (20.56 μg/kg) and Cumilla (11.91 μg/kg) revealing more aflatoxins contamination than the regulatory limit (10 μg/kg) in which only aflatoxins concentration from Kushtia and Kishoreganj revealed statistically significant data, besides, rest of the location exhibited below level of aflatoxins contamination of regulatory limit showing more or less statistically significant data.

Total aflatoxins associated with maize grains were detected in 2020, with the supreme value was detected in Chuadanga (30.5 %) followed by Kushtia (29.5 %), Nilphamari (22.5 %), Panchagarh (19.25 %) and the minimal aflatoxins was detected in Manikganj (3.2 %), rest of the samples from other districts revealed lower to moderate level of aflatoxins, moreover, data from Chuadanga and Kushtia, Cumilla, Jashore and Natore, Thakurgaon and Rangpur, Lalmonirhat and Kishoreganj regions revealed ststistically similar data while data from other regions exhibited statistically dissimilar data.

In case of infection rate, toxgenic maize samples were obtained from Panchagarh, Thakurgoan, Gaibandha, Chuadanga, Kishoreganj exhibiting 100 % infection by

A. flavus and no atoxigenic samples were found in those area. Moderate amount of toxigenic *A. flavus* was detected in Jashore followed by Cumilla, Natore, Lalmonirhat, Nilphamari which were 78 %, 75 %, 66 %, 50 % respectively and atoxigenic fungi was detected 22 %, 25 %, 34 %, 50 %, 50 % were detected respectively. Rest of the locations (Dinajpur, Rangpur, Bogura, Kushtia, Manikganj) exhibited higher amount of atoxigenic *A. flavus* compared to toxigenic *A. flavus* (**Table 1**).

The outmost percent aflatoxins concentration over standard limit was found in Chuadanga (915.7 %) followed by Gaibandha (587.3 %), Kushtia (214.8 %), Kishoreganj (208.85 %), Rangpur (105.6 %), Cumilla (19.5 %) revealing that the aflatoxins contamination from those area were beyond the regulatory limit set by EU for aflatoxins (10 ppb), conversely, aflatoxin concentration from other nine locations were below the regulatory limit of aflatoxins (**Table 1**).

8.2 Relationship between aflatoxins producing *A. flavus* and mean aflatoxins concentrations

The regression analysis between toxigenic *A. flavus* percentage and mean aflatoxin concentrations which was positively correlated by observing regression equation where the slope was = 0.55 and y-intercept was = 50.14, coefficient of

Location	Total aflatoxins concentrations (μg/kg)	% *A. flavus* associated with maize grains			Percent total aflatoxins concentration over standard limit
		Total	Toxigenic	Atoxigenic	
Panchagarh	4.96 ± 0.19^{f}	19.25 ± 3.53^{c}	100	0	—
Thakurgoan	1.28 ± 0.10^{g}	18.25 ± 0.43^{cd}	100	0	—
Dinajpur	1.08 ± 0.122^{g}	16 ± 1.73^{de}	25	75	—
Nilphamari	3.04 ± 0.56^{fg}	22.5 ± 3.28^{b}	50	50	—
Rangpur	20.56 ± 0.42^{d}	18.5 ± 2.18^{cd}	44	56	105.6
Lalmonirhat	3.37 ± 0.19^{fg}	9.75 ± 1.00^{g}	50	50	_
Gaibandha	68.73 ± 4.02^{b}	3.75 ± 1.00^{h}	100	0	587.3
Bogura	3.33 ± 0.41^{fg}	11.25 ± 0.66^{fg}	40	60	—
Natore	2.39 ± 1.29^{fg}	13.5 ± 1.80^{ef}	66	34	—
Kushtia	31.48 ± 1.14^{c}	29.5 ± 1.32^{a}	33	67	214.8
Jashore	1.67 ± 0.57^{g}	13.75 ± 1.64^{ef}	78	22	—
Chuadanga	101.57 ± 5.09^{a}	30.5 ± 0.50^{a}	100	0	915.7
Kishorerganj	30.89 ± 0.22^{c}	10.25 ± 1.09^{g}	100	0	208.85
Manikganj	2.57 ± 0.01^{fg}	3.25 ± 0.43^{h}	33.33	66.67	—
Cumilla	11.91 ± 0.30^{e}	14 ± 2.00^{ef}	75	25	19.5
Level of significance	**	**			
LSD	2.07	2.95			
CV	5.07	9.66			

**Significant at 5 % level of significance. Least significant difference (LSD) at P = 0.05 was used for comparing means and the P values were 0.00.*
***Significant at 1 % level of significance. Least significant difference (LSD) at P = 0.05 was used for comparing means and the P values wee 0.00. Data are the averages of three biological replications. The regulatory limits for fumonisin is 1 ppm (10 μg/kg).*

Table 1.
Levels of Total aflatoxins concentration in stored maize grains collected from the stores of traders of fifteen maize growing areas of Bangladesh.

determination, R^2 = 0.198 and coefficient of correlation, r = 0.44 which depicted that 1 % surges of toxigenic *A. flavus* in maize grains ultimately rised 50.137 μg/kg aflatoxin concentration. In terms of 5 % surges of toxigenic *A. flavus* in maize grains, the aflatoxin concentration was increased up to 2.75 μg/kg and when toxigenic *A. flavus* increased 20 % in maize grains, the aflatoxin concentration was escalated up to 11.0 μg/kg (**Figure 1**).

8.3 Identification of *A. flavus* from the stored maize grain samples collected from some selected growing areas of Bangladesh

Morphological identification of *A.flavus* was detected by using petridish and culture plate method as well as observing microscopic figures under compound and stereo microscope (**Figure 2A(a)–(d)**). Thirty five fungal isolates were identified using primers specific to ITS 1 and ITS 4 regions. PCR assays of *A. flavus* DNA with ITS 1 and ITS 4 primers amplified a single fragment of about 600 bp which revealed that all the isolates obtained were fungi. Sequence analysis of ITS region by BLAST program revealed that all the isolates obtained from maize were belong to *A. flavus* (**Figure 3A**).

8.4 PCR based identification and confirmation of aflatoxin producing *Aspergillus flavus* species obtained from maize grain samples

AF02_Ran, AF01_Lal, AF01_Bog, AF02_Bog, AF03_Jas, AF04_Jas, AF01_Chu, AF03_Kis, AF04_Kis, AF01_Man were identified by PCR amplification of ITS region using ITS1 and ITS4 primers and the results of PCR showed an amplification size 600 bp confirmed the fungal isolates (**Figure 3A**) and their several strains were found in Rangpur (*A. flavus* strain 64-A1), Lalmonirhat (*A. flavus* strain SGE22), Bogura (*A. flavus* strain SGE22 and *A. flavus* strain bpo4), Jashore (*A. flavus* and *A. flavus* isolate AA221), Chuadanga (*A. flavus* strain JN-YG-3-5), Kishoreganj (*A. flavus* strain 64-A1 and *A. flavus* strain ND26), Manikganj (*A. flavus* strain SU-16).

PCR products were then sequenced and analysis of sequence data of amplified ITS region using BLAST program revealed that fungal isolates AF01_Man, AF03_Jas, AF02_Ran obtained from maize grain samples collected from Manikganj, Jashore, Rangpur revealed the highest homology of 99.33 %, 99.17 %, 95.74 % with the *A. flavus strain* SU-16, *A. flavus, A. flavus strain* 64-A1. Other sevel isolates obtained from Lalmonirhat (AF01_Lal), Bogura (AF01_Bog), Bogura (AF02_Bog), Jashore (AF04_Jes), Chuadanga (AF01_Chu), Kishoreganj (AF03_Kis), Kishoreganj (AF04_Kis) showed significant homology with different strains of *A. flavus* (**Table 2**).

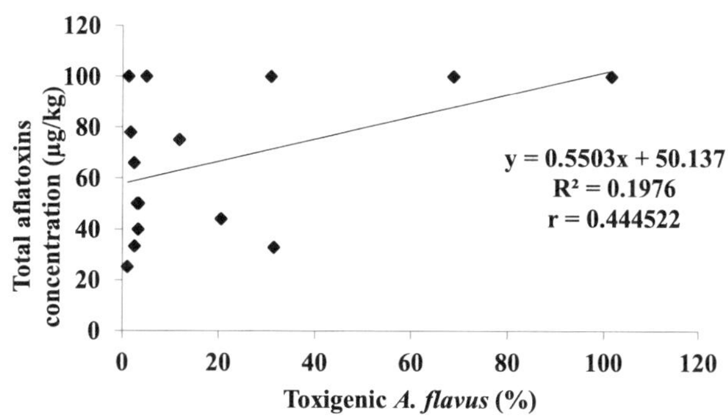

Figure 1.
Linear correlations between toxigenic A. flavus *infected maize grains and total aflatoxins concentration.*

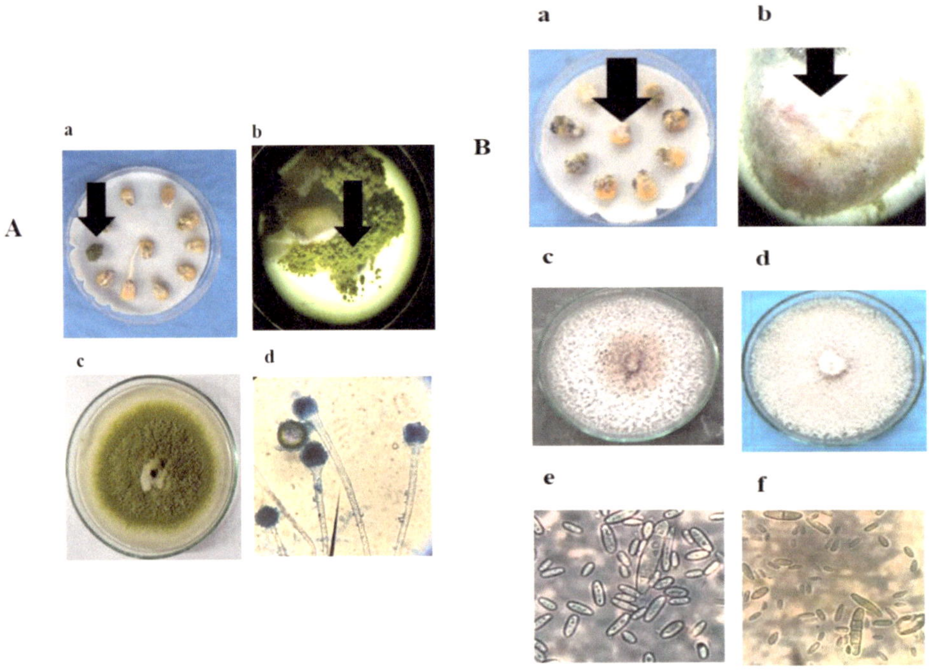

Figure 2.
(A) Composite photographs of Aspergillus *spp. in different sections. (a) Apparent growth of* Aspergillus *spp. on the maize grain surface, (b) enlarged view of individual maize grain showing the growth of* Aspergillus *spp., morphology of suspected* Aspergillus *spp. (c) Yellowish green colonies of* A. flavus *on PDA, (d) vesicle with less conidial ornamentation with conidiphores of* A. flavus. *(B) Composite photographs of* Fusarium *spp. in different sections. (a) Apparent growth of* Fusarium *spp. on the maize grain surface, (b) enlarged view of individual maize grain showing the growth of* Fusarium *spp., morphology of suspected* Fusarium *spp., (c) pinkish white growth of* F. proliferatum *on PDA, (d) microconidia of* F. proliferatum *without septum under microscope with 40× magnification, (e) whitish growth of* F. oxysporum *on PDA and (f) Micro and macro-conidia (with septum) of* F. oxyporum *without septum. Culture photographs were taken at 7 days after inoculation and microscopic photographs were taken with 40× magnification using compound light microscope equipped with a digital camera.*

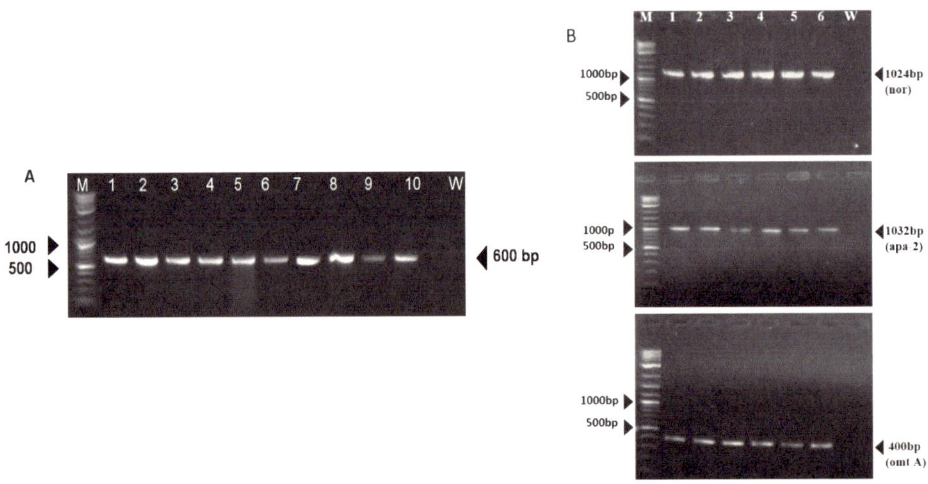

Figure 3.
(A) PCR amplification of ITS region from the genomic DNA of the fungal isolates using ITS-1 and ITS-4 primers and (B) PCR amplification of nor, omt, apa-2 gene from the genomic DNA of the fungal isolates obtained from obtained from fifteen maize growing areas of Bangladesh M: 1 kb plus DNA ladder, 1, AF02_Ran: Rangpur, 2, AF01_Lal: Lalmonirhat, 3, AF01_Bog: Bogura, 4, AF02_Bog: Bogura, 5, AF03_Jas: Jassore, 6, AF04_Jas: Jashore, 7, AF01_Chu: Chuadanga, 8, AF03_Kis: Kishoreganj, 9, AF04_Kis:Kishoreganj, 10, AF01_Man: Manikgan.

Isolate ID	Location	Closest relatives	Accession numbers	Identity	Homology (%)	Aflatoxins biosysthesis genes			Comment
						nor	Omt A	apa	
AF01_Pan	Panchagarh	*A. flavus* isolate PA223	MN006634.1	422/428	98.6	—	+	+	Toxigenic
AF02_Pan	Panchagarh	*A. flavus* strain AF15	KX253943.1	194/204	95.1	+	—	—	Toxigenic
AF01_Tha	Thakurgoan	*A. flavus* strain SU-16	MT680400.1	95/99	95.96	+	—	—	Toxigenic
AF02_Tha	Thakurgoan	*A. flavus* isolate AA221	MN006401.1	171/178	96.07	+	—	—	Toxigenic
AF01_Din	Dinajpur	*A. flavus* isolate 2011F7	MT558941.1	595/598	99.5	+	+	—	Toxigenic
AF01_Nil	Nilphamari	*A. flavus* isolate Z15	MH237650.1	88/90	97.78	+	—	—	Toxigenic
AF02_Nil	Nilphamari	*A. flavus* strain SGE34	JQ776536.1	505/522	96.74	+	+	—	Toxigenic
AF01_Ran	Rangpur	*A. flavus* strain SU-16	MT680400.1	95/99	95.96	+	—	—	Toxigenic
AF02_Ran	Rangpur	*A. flavus* strain 64-A1	MT594359.1	90/94	95.74	+	+	+	Toxigenic
AF03_Ran	Rangpur	*A. flavus* strain SU-16	MT680400.1	416/427	97.42	—	+	—	Toxigenic
AF04_Ran	Rangpur	*A. flavus* strain 64-A1	MT594359.1	474/497	95.37	+	—	—	Toxigenic
AF01_lal	Lalmonirhat	*A. flavus* strain SGE22	JX232269.1	333/370	90	+	+	+	Toxigenic
AF01_Gai	Gaibandha	*A. flavus* isolate A3	MH237624.1	71/72	98.61	+	—	—	Toxigenic
AF01_Bog	Bogura	*A. flavus* strain SGE22	JX232269.1	403/446	90.36	+	+	+	Toxigenic
AF02_Bog	Bogura	*A. flavus* strain bpo4	MT492458.1	424/449	94.43	+	+	+	Toxigenic
AF03_Nat	Natore	*A. flavus* strain BLND1-1	MN396712.1	400/428	93.46	—	+	—	Toxigenic
AF01_Nat	Natore	*A. flavus* strain GFRS16	MT447484.1	591/608	97.2	+	—	—	Toxigenic
AF01_Kus	Kushtia	*A. flavus*solate V5F-13	HQ395774.1	310/321	96.57	+	—	—	Toxigenic
AF01_Jas	Jashore	*A. flavus* isolate BB-1	MT584825.1	577/600	96	+	—	—	Toxigenic
AF02_Jas	Jashore	*A. flavus* isolate AA221	MN006401.1	72/73	98.63	+	—	—	Toxigenic
AF03_Jas	Jashore	*A. flavus*	MN238861.1	599/604	99.17	+	+	+	Toxigenic

Isolate ID	Location	Closest relatives	Accession numbers	Identity	Homology (%)	Aflatoxins biosysthesis genes			Comment
						nor	Omt A	apa	
AF04_Jas	Jashore	*A. flavus* isolate AA221	MN006401.1	229/241	95	+	+	+	Toxigenic
AF05_Jas	Jashore	*A. flavus* strain BLND1-1	MN396712.1	157/164	95.73	+	—	—	Toxigenic
AF06_Jas	Jashore	*A. flavus* strain A1	CP051065.1	551/587	93.87	+	—	+	Toxigenic
AF07_Chu	Jashore	*A. flavus* strain FG38	EU030347.1	38/39	97.44	+	—	—	Toxigenic
AF01_Chu	Chuadanga	*A. flavus* strain JN-YG-3-5	MG554231.1	413/457	90.37	+	+	+	Toxigenic
AF01_Kis	Kishoreganj	*A. flavus* isolate AA221	MN006401.1	469/480	97.71	+	—	—	Toxigenic
AF01_Kis	Kishoreganj	*A. flavus* strain 64-A1	MT594359.1	144/150	96	+	—	+	Toxigenic
AF02_Kis	Kishoreganj	*A. flavus* strain JN-YG-3-5	MG554231.1	412/455	90.55	+	+	—	Toxigenic
AF03_Kis	Kishoreganj	*A. flavus* strain 64-A1	MT594359.1	146/154	94.81	+	+	+	Toxigenic
AF04_Kis	Kishoreganj	*A. flavus* strain ND26	MG659620.1	384/443	86.68	+	+	+	Toxigenic
AF01_Man	Manikganj	*A. flavus* strain SU-16	MT680400.1	591/595	99.33	+	+	+	Toxigenic
AF01_Cum	Cumilla	*A. flavus* isolate PA223	MN006634.1	304/317	95.9	+	—	—	Toxigenic
AF02_Cum	Cumilla	*A. flavus* strain train YLF-14	HQ400610.1	63/67	94.03	+	+	—	Toxigenic
AF03_Cum	Cumilla	*A. flavus* strain JN-YG-3-5	MG554231.1	304/358	85	—	+	—	Toxigenic

Table 2.
List of A. flavus *isolates identified by homology search of sequences of ITS region by BLAST program obtained from maize grain samples collected from fifteen growing areas of Bangladesh.*

When the isolates of *Aspergillus* Spp. were analyzed by PCR for aflatoxin producing ability using *nor*, *omtA*, *apa-2* genes based primers from fifteen maize growing areas. The result showed the amplified DNA fragment was 400 bp, 1024 bp, 1032 bp confirmed that the *A. flavus* isolates had the ability to produce aflatoxin that encode *nor*, *omtA*, *apa-2* genes (**Figure 3B**). Only six species showed a positive result with *nor*, *omtA*, *apa-2* genes set of primers. The result indicated *A. flavus* strains were aflatoxins producers as those were an evident from our investigation (**Figure 3B**).

PCR products were sequenced using ITS-1 primer and sequence data were analyzed by homology search using BLAST Nucleotide program. Isolates were identified as different *A. flavus* based on the homology percentage with their closest relatives available in the NCBI database.

9. Determination of total fumonisins contamination in stored maize grain samples collected from some selected growing areas of Bangladesh

The study was conducted at the Laboratory of Department of Plant Pathology, Bangladesh Agricultural University, Mymensingh. Composite stored maize grains samples were collected from 15 maize growing areas of Bangladesh such as Panchagarh, Thakurgaon, Dinajpur, Nilphamari, Rangpur, Lalmonirhat, Gaibandha, Bogura, Natore, Kushtia, Jashore, Chuadanga, Kishoreganj, Manikganj, Cumilla.

Fumonisins were detected with the highest value recorded in Gaibandha (9.18 mg/kg) and the lowest in Cumilla (0.11 mg/kg) (**Table 3**). Panchagarh (1.47 mg/kg), Thakurgaon (1.27 mg/kg), Dinajpur (0.65 mg/kg), Nilphamari (1.28 mg/kg), Rangpur (1.65 mg/kg), Lalmonirhat (1.18 mg/kg), Bogura (1.29 mg/kg), Kushtia (1.44 mg/kg), Kishoreganj (1.54 mg/kg), and Manikganj (1.47 mg/kg) had moderately high fumonisin levels revealing statistically identical data. Other regions showed indentically dissimilar data except Natore (0.23 mg/kg) and Chuadanga (0.59 mg/kg) (**Table 3**).

Infection rate of *Fusarium* spp. had the highest value in Bogura (13.50 %) followed by Gaibandha (13.25 %), Nilphamari (12.50 %) depicted statistically similar data and the minimal was found in Chuadanga (0.50 %) and Kustia (0.56 %). Moderately higher levels of fumonisin detected in Panchagarh (2.63 %), Thakurgaon (6.06 %), Dinajpur (2.38 %), Rangpur (9.69 %), Jessore (2.25 %), Kishoreganj (17.88 %), Manikganj (6.94 %) and Cumilla (5.31 %) were in the group of ststistically identical data. Moderate but less high and statistically similar results showed in Thakurgaon (6.06 %) and Manikganj (6.94 %) (**Table 3**).

The outmost percent fumonisins concentration over standard limit was found in Rangpur (65 %) followed by Kishoreganj (53.5 %), Gaibandha (47.5 %), Manikganj (47 %), Kushtia (45 %), Panchagarh (46.5 %), Bogura (28.5 %), Thakurgaon (27 %), Nilphamari (27 %), Lalmonirhat (18 %) revealing that the aflatoxins contamination from those area were beyond the regulatory limit set by EU for fumonisins (1 ppm), conversely, fumonisins concentration from other five locations were below the regulatory limit of fumonisins (1 ppm) (**Table 3**).

9.1 Relationship between fumonisins producing *Fusarium* spp. and mean fumonisin concentrations

The regression analysis between *Fusarium* spp. infected maize grains and mean fumonisin concentrations which was positively correlated by observing regression equation where the slope was = 0.038 and y-intercept was = 0.882, coefficient of

Location	Total fumonisins (mg/kg)	Percent maize grains infected with *Fusarium* species	Percent total Fumonisins concentration over standard limit
Panchagarh	1.47 ± 0.14^{b}	2.63 ± 1.20^{e}	46.5
Thakurgoan	1.27 ± 0.13^{b}	6.06 ± 2.07^{cd}	27
Dinajpur	0.65 ± 0.01^{d}	2.38 ± 0.54^{ef}	—
Nilphamari	1.28 ± 0.11^{b}	12.50 ± 0.89^{a}	27
Rangpur	1.65 ± 0.27^{b}	9.69 ± 2.33^{b}	65
Lalmonirhat	1.18 ± 0.17^{bc}	0.00 ± 0.00^{h}	18
Gaibandha	9.18 ± 1.02^{a}	13.25 ± 1.39^{a}	47.5
Bogura	1.28 ± 0.33^{b}	13.50 ± 1.5^{a}	28.5
Natore	0.23 ± 0.06^{de}	0.00 ± 0.00^{h}	—
Kushtia	1.44 ± 0.1^{b}	0.56 ± 0.41^{fgh}	45
Jashore	0.75 ± 0.10^{cd}	2.25 ± 0.43^{efg}	—
Chuadanga	0.59 ± 0.07^{de}	0.50 ± 0.50^{gh}	—
Kishoreganj	1.54 ± 0.20^{b}	7.88 ± 0.82^{bc}	53.5
Manikganj	1.47 ± 0.22^{b}	6.94 ± 0.91^{cd}	47
Cumilla	0.11 ± 0.01^{e}	5.31 ± 0.35^{d}	—
Level of significance	**	**	
LSD	0.52	1.86	
CV (%)	12.99	15.97	

**Significant at 5% level of significance. Least significant difference (LSD) at P = 0.05 was used for comparing means and the P values were 0.00.*
***Significant at 1% level of significance. Least significant difference (LSD) at P = 0.05 was used for comparing means and the P values were 0.00. Data are the averages of three biological replications. The regulatory limits for fumonisin is 1 ppm (1 mg/kg).*

Table 3.
Levels of total fumonisins concentration in stored maize grains collected from the stores of traders of fifteen maize growing areas of Bangladesh.

determination, $R^2 = 0.198$ and coefficient of correlation, r = 0.45 which depicted that 1 percent surges of *Fusarium* in maize grains ultimately rised 0.038 mg/kg fumonisins concentration. In terms of 5 % surges of *Fusarium* in maize grains, the fumonisins concentration was increased up to 0.19 mg/kg and when *Fusarium* increased 20 % in maize grains, the fumonisins concentration was escalated up to 0.76 mg/kg (**Figure 4**).

9.2 Identification of *Fusarium* species from the stored maize grain samples collected from some selected growing areas of Bangladesh

Morphological identification of *F. oxysporum* and *F. proliferatum* were detected by using petridish and culture plate method as well as observing microscopic figures under compound and stereo microscope (**Figure 2B(a)-(f)**). Fifteen fungal isolates were identified using primers specific to ITS 1 and ITS 4 region. PCR assays of *F. oxysporum* DNA with ITS 1 and ITS 4 primers amplified a single fragment of about 600 bp which revealed that all the isolates obtained were fungi (**Figure 5A**). Sequence analysis of ITS region by BLAST program revealed that all the isolates obtained from maize were belong to *F. oxysporum* and *F. proliferatum*.

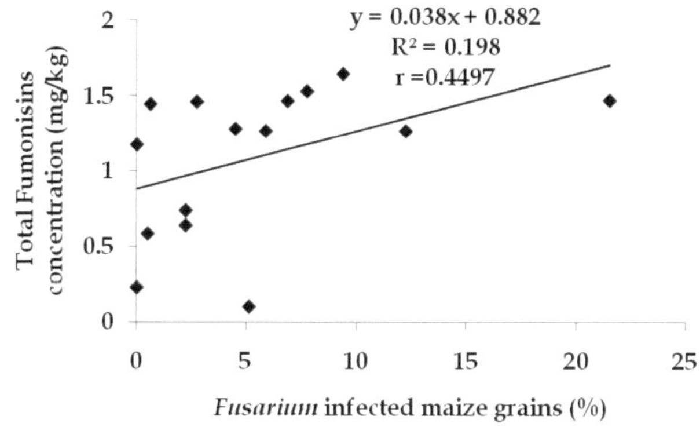

Figure 4.
Linear correlations between Fusarium *infected maize grains and total fumonisin concentration.*

10. PCR based identification and confirmation of fumonisins producing *Fusarium* species obtained from maize grain samples

F01_Pan, F02_Tha, F03_Din, F04_Nil, F05_Ran, F06_Lal, F07_Gai, F08_Bog, F09_Nat, F010_Kus, F011_Jes, F012_Chu, F013_Kis, F014_Man and F015_Cum were identified by PCR amplification of ITS region using ITS1 and ITS4 primers and the results of PCR showed an amplification size 600 bp confirmed the *Fusarium*. PCR products were then sequenced. (**Figure 2A**). Out of fifteen maize growing areas, *F. oxysporum* was found in Panchagarh (*F. oxysporum* strain EP19), Thakurgaon (*F. oxysporum* strain En3), Dinajpur (*F. oxysporum* strain EP19), Nilphamari (*F. oxysporum* strain EP19), Rangpur (*F. oxysporum* strain En3), Natore

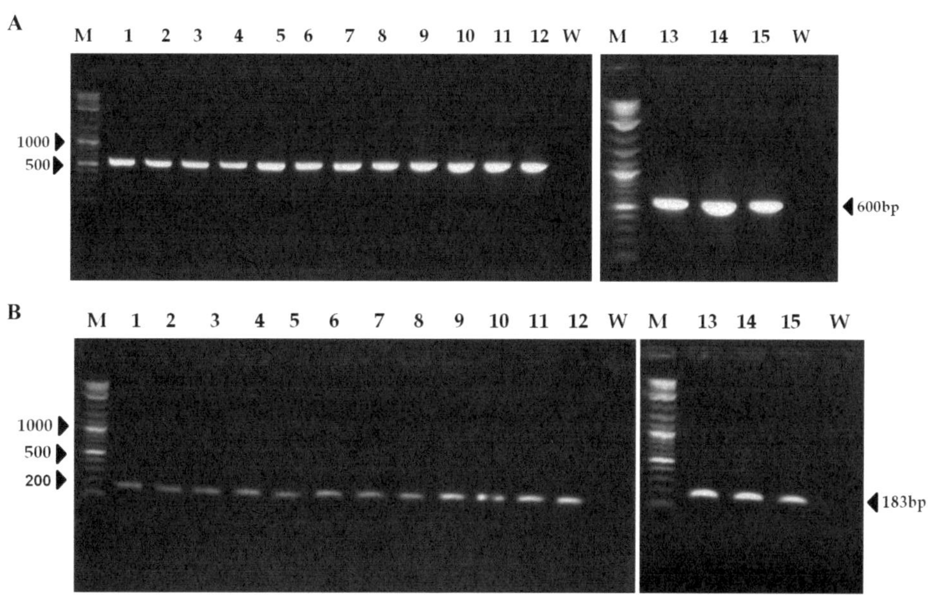

Figure 5.
A. PCR amplification of ITS region from the genomic DNA of the fungal isolates using ITS-1 and ITS-4 primers and B. PCR amplification of FUM1 gene from the genomic DNA of the fungal isolates obtained from obtained from fifteen maize growing areas of Bangladesh M: 1 kb plus DNA ladder, 1, F01_Pan: Panchagarh, 2, F02_Tha: Thakurgoan, 3, F03_Din: Dinajpur, 4, F04_Nil: Nilphamari, 5, F05_Ran: Rangpur, 6, F06_Lal: Lalmonirhat, 7, F07_Gai: Gaibandha, 8, F08_Bog: Bogura, 9, F09_Nat: Natore, 10, F010_Kus: Kushtia, 11, F011_Jes: Jashore, 12, F012_Chu: Chuadanga, 13, F013_Kis: Kishoreganj, 14, F014_Man: Manikganj and 15, F015_Cum:Cumilla.

(*F. oxysporum* isolate FH10 18S), Kushtia (*F. oxysporum* strain EP19), Jashore (*F. oxysporum* strain En3), Chuadanga (*F. oxysporum* isolate H200714-017) Manikganj (*F. oxysporum* strain EP19), Cumilla (*F. oxysporum* strain En3) and *F. proliferatum* was found in Lalmonirhat (*F. proliferatum* strain TH11-3), Gaibandha (*F. proliferatum* strain TH11-3), Bogura (*F. proliferatum* strain TH11-3) and Kishoreganj (*F. proliferatum* strain TH11-3).

Fungal isolates F06_Lal, F07_Gai, F08_Bog and F013_Kis obtained from maize grain samples were collected from Lalmonirhat, Gaibandha, Bogura and Kishoreganj showed the highest homology with *F. proliferatum* strain TH11-3 (**Table 4**). The fungal isolates obtained from maize grain samples collected from

Isolate ID	Location	Closest relatives	Accession number	Identity	Homology (%)
F01_Pan	Panchagarh	*F. oxysporum* strain EP19	MN704852.1	486/534	91.01
F02_Tha	Thakurgoan	*F. oxysporum* strain En3	MN726603.1	491/537	91.43
F03_Din	Dinajpur	*F. oxysporum* strain EP19	MN704852.1	445/530	83.96
F04_Nil	Nilphamari	*F. oxysporum* strain EP19	MN704852.1	486/534	91.01
F05_Ran	Rangpur	*F. oxysporum* strain En3	MN726603.1	477/539	88
F06_Lal	Lalmonirhat	*Fusarium proliferatum* strain TH11-3	MT563411.1	472/508	92.91
F07_Gai	Gaibandha	*Fusarium proliferatum* strain TH11-3	MT563411.1	472/508	92.91
F08_Bog	Bogura	*Fusarium proliferatum* strain TH11-3	MT563411.1	491/544	90
F09_Nat	Natore	*F. oxysporum* isolate FH10 18S	KU361495.1	257/305	84.26
F010_Kus	Kushtia	*F. oxysporum* strain EP19	MN704852.1	486/534	91.01
F011_Jes	Jashore	*F. oxysporum* strain En3	MN726603.1	477/539	88
F012_Chu	Chuadanga	*F. oxysporum* isolate H200714-017	MT974426.1	477/541	88.17
F013_Kis	Kishoreganj	*F. oxysporum* strain TH11-3	MT563411.1	472/508	92.91
F014_Man	Manikganj	*F. oxysporum* strain EP19	MN704852.1	486/534	91.01
F015_Cum	Cumilla	*F. oxysporum* strain En3	MN726603.1	477/539	88

PCR products were sequenced using ITS-1 primer and sequence data were analyzed by homology search using BLAST Nucleotide program. Isolates were identified as different Fusarium species based on the homology percentage with their closest relatives available in the NCBI database. F01_Pan: Panchagarh, F02_Tha: Thakurgoan, F03_Din: Dinajpur, 4, F04_Nil: Nilphamari, F05_Ran: Rangpur, F06_Lal: Lalmonirhat, F07_Gai: Gaibandha, F08_Bog: Bogura, F09_Nat: Natore, F010_Kus: Kushtia, F011_Jes: Jashore, F012_Chu: Chuadanga, F013_Kis: Kishoreganj, F014_Man: Manikganj and F015_Cum: Cumilla.

Table 4.
List of Fusarium isolates identified by homology search of sequences of ITS region by BLAST program obtained from maize grain samples collected from fifteen growing areas of Bangladesh.

Panchagarh (F01_Pan), Thakurgaon (F02_Tha), Dinajpur (F03_Din), Nilphamari (F04_Nil), Rangpur (F05_Ran), Lalmonirhat (F06_Lal), Gaibandha (F07_Gai), Bogura (F08_Bog), Natore (F09_Nat), Kustia (F010_Kus), Jessore (F011_Jes), Chuadanga (F012_Chu), Kishoreganj (F013_Kis), Manikganj (F014_Man) and Cumilla (F015_Cum) showed significant homology with different strains of *F. oxysporum* (**Table 4**).

When the isolates of *Fusarium* species were analyzed by PCR for fumonisins producing ability using *FUM1* gene based primers from fifteen maize growing areas. The result showed the amplified DNA fragment was 183 bp confirmed that the *Fusarium* had the ability to produce fumonisin that encode *FUM1* gene (**Figure 5B**). Only two *Fusarium* species showed a positive result with *FUM1* gene set of primers. The result was contrary as *F. proliferatum* and *F. oxysporum* (**Table 4**) were fumonisin-producers as it was evident from our investigation.

11. Discussion

The experiment was conducted at Plant Bacteriology and Biotechnology Laboratory of Department of Plant Pathology, Bangladesh Agricultural University, Mymensingh during the period of 2019–2020. The purpose of the experiment were to detect the levels of fumonisins and aflatoxins and to identify the aflatoxin and fumonisins producing *Aspergillus* and *Fusarium* in maize associated with maize by PCR using nor, omtA, apa-2 and *FUM1*. Genes involving *afl R*, *ver-1*, *omt-1* and *apa-2* associated with biosynthetic pathway regarding aflatoxins production [73–76]. *Apa-1*, *Nor-1*, *Omt-1* and *Ver-1* gens belong to four primers were applied to detect aflatoxins contamination [77, 78]. *A. flavus* was quantified by *nor-1* gene in several contaminated food samples and cereals using PCR assay [77]. Besides, [56] mentioned that *FUM1* gene with an expected amplicon size of 183 bp can easily detect the fumonisin and non-fumonisin producing *Fusarium,* moreover other researchers also identified the fumonisin by using *FUM1* gene which is in accordance with our study [79–81]. We gathered samples from 15 maize growing areas to measure the aflatoxins and fumonisins level but not all the *Aspergillus* strains are capable of engendering mycotoxins, thus screening is crucial and we detected by Agra Quant Total Aflatoxin and Fumonisin Test Kit following ELISA approach for detection and this method also used by [82–87] for detecting aflatoxins and fumonisin. In our experiment, we detected the aflatoxins contamination Agra Quant Total Aflatoxins 96 well microtiter plate ELISA test kit produced in Romer Labs, Packers and Stockyards Administration (GIPSA) in US Department of Agriculture (USDA) which ability to detect individual aflatoxins very precisely and accurately with a range of 0–320 ppb in accrodance with an experiment conducted by [82]. A number of approaches have been widely used to detect mycotxin naming high-performance liquid chromatography (HPLC), enzyme-linked immunosorbent assay (ELISA), and thin layer chromatography (TLC) [83, 84] and served as a reliable method for detecting aflatoxins and fumonisins [85, 88, 89]. In Gaibandha and Cumilla region fumonisin contamination were highest and lowest compared to other areas revealing moderate amount of fumonisins. In this study, all of the 15 samples were found positive with fumonisins producing *Fusarium* and aflatoxin producing fungi *Aspergillus* which in accordance with the findings of [90, 91]. We found positive correlation for both aflatoxins and fumonisins contamination between their toxin percentages which were matched with the findings of [92] who found apositive correlation has been identified between the proportion of *FUM1* transcripts and the proportion of fumonisins biosynthesized by the *F. verticillioides* and *F. proliferatum* species.

In case of Percent total Fumonisins concentration over standard limit, five regions were under the regulatory limit and other ten regions were exposed higher limit than the regulatory limit exhibiting 65 % followed by 53.5 %, 47.5 %, 47 %, 46.5 %, 45 %, 28.5 %, 27 %, 27 %, 18 % over the standard limit (1 ppm) in the area of Rangpur, Kishoreganj, Gaibandha, Manikganj, Panchagarh, Kustia, Bogura, Nilphamari, Thakurgaon, Lalmonirhat respectively. On the other hand, highest and lowest aflatoxin concentration was recorded in Chuadanga and Dinajpur regions and in terms of percent aflatoxin concentration over standard limit, eight regions were below the permissible limit of aflatoxins, conversely, five regions exposing 915.7 % followed by 587.3 %, 214.8 %, 208.85 %, 19.5 % aflatoxin concentration beyond permissible limit of 10 μg in the region of Chuadanga, Gaibandha, Kustia, Kishoreganj and Cumilla respectively. Refs. [15, 93] recorded that surges of aflatoxin contamination levels beyond regulatory limit due to increased droughts, pest damages, temperatures, host susceptibility.

As we observed that both aflatoxin and fumonisin concentration were fluctuate one region to another region which have been also monitored that due to association of several significant factors like temperature, water activity, storage conditions, drought, humidity, insect damage, flowering stage, plant characteristics [94–98]. Ref. [48] revealed that aflatoxin production comprised of several factor including existence of certain genes and in intact that means deletions or insertions within the gene regions, crop stress [99] and in fumonisins two factors temperatures and water potential are fundamental to produce fumonisins [99] along with rainfall patterns, longer durations of drought which has been prominent in Mediterranean regions [100–103]. These all conditions significantly impact on the variation of the population of mycotoxin producing fungi both *Fusarium* and *Aspergillus* [103]. In our experiment, we recorded over all three regions (Chuadanga, Kishoreganj, Gaibandha) were engendering higher amount of aflatoxins and fumonisins production respectively, thus we speculated in Chuadanga, temperature fluctuation influences the mycotoxin production, in Kishoreganj which exposed with flood and severe water stress and the region Gaibandha with drought problems, these might have the feasible factor for *Aspergillus* and *Fumonisins* to produce gigantic amount of mycotoxins compared to other areas. Aflatoxin levels rise as a result of drought, insect damage, and heat during fungal growth [25]. Marasas [104] found that, the presence of fumonisins is linked to weather conditions, with larger instances occurring during hot and dry conditions. Abbas et al. [105] revealed that *A. flavus* grows supreme around 28–37° C with a humidity level of at least 80 %.

Post-harvest factors are also exacerbate mycotoxin production and generate a favorable condition for fungus related to their growth and mycotoxin production and those include storage fungus, insect infestation, contaminant mold respiration, insects and mites, water availability and temperature ultimately deteriorate grain quality [106–108]. As [109] also observed that interaction between these factors triggered the mycotoxigenic species growth, mycotoxin production, niche occupation and competitiveness, [110] also revealed the moisture and surrounding air conditions also influenced mycotoxin production by initiating biological and biochemical activity. Maize is a hygroscopic crop which easily absorbs or release moisture and humidity in the surrounding ambience until getting the adjustment with equilibrium conditions which led to swift degradation in storage. *Fusarium* species can damage stored grain by causing seedling illnesses, root rots, stalk rots, and ear rots in maize which ultimately hazardous to plants and animal [111–116]. Due to all correlating factors with aflatoxin production, high amount of aflatoxins were found in Bangladeshi markets [23] and 82 % contamination in South Asia [49]. Decomposing potentiality of AFs are very slow several approaches including physical, chemical have been investigated [19] and monitored changing in sensory property and nutrient diminishment which led to mount food safety problems ultimately. A number of microorganisms have been identified fruitfully working as a biocontrol agents to control mycotoxins such as *Bacillus subtilis*, *Pseudomonas*, *Trichoderma*, atoxigenic strains of *A. flavus* and *A. parasiticus* [117–119]. Thus, suppressing mycotoxins by biocontrol agent would be a fruitful approach though several experiments need to be conducted precisely in future.

12. Conclusion

Aflatoxins and fumonisins are the major source of disease outbreaks due to a lack of knowledge and consumption of contaminated food and feed in Bangladesh. Excessive levels of aflatoxins and fumonisins in food in Bangladesh is a major concern because still majority of the people have not any idea that they are consuming food and feed which crossed the permissible limit set by EU. Another significant factor is no sign of regulating any acceptable limit for this country and that's why people are easily contaminated with several mycotoxins without properly knowing any acceptable limit as well as industries are also not ensuring any precise step to diminish mycotoxins concentration in terms of engendering several products. As our study clearly conceded that most of the regions (Rangpur, Gaibandha, Kushtia, Chuadanga, Kishoreganj, Manikganj, Cumilla) were at higher risk for aflatoxin as well as the regions (Panchagarh, Thakurgoan, Nilphamari, Rangpur, Lalmonirhat, Gaibandha, Bogura, Kushtia, Kishoreganj, Manikganj) were exposed with fumonisins contamination more than that of acceptable limit of fumonisins which ultimately effects animal and mankind by entering our food chain. Thus, several effective approaches (physical, chemical, biological, and genetic engineering techniques) need to be employed as early as possible to suppress the ruinous consequences of mycotoxin contamination of Bangladesh.

Acknowledgements

This research work was financed by BAURES (Bangladesh Agricultural University Research System) to Dr. Md. Rashidul Islam (Grant no.: 2020/972/BAU), Department of Plant Pathology, Bangladesh Agricultural University, Mymensingh-2202 as well as from Bangabandhu Science and Technology Fellowship Trust, Ministry of Science and Technology to Muhtarima Jannat.

Conflict of interest disclosure

Authors do not have any conflict of interests to declare.

Author details

Muhtarima Jannat[1], Md. Mostafa Masud[1], Mushfika Nusrat[1], Samrin Bashar[1], Mamuna Mahjabin Mita[1], Muhammad Iqbal Hossain[1], Md. Zahangir Alam[1], Sabina Yeasmin[2] and Md. Rashidul Islam[1*]

1 Laboratory of Plant Bacteriology and Biotechnology, Department of Plant Pathology, Bangladesh Agricultural University, Mymensingh, Bangladesh

2 Agro Innovation Laboratory, Department of Agronomy, Bangladesh Agricultural University, Mymensingh, Bangladesh

*Address all correspondence to: rashidul.islam@bau.edu.bd

References

[1] Aldrich SR, Scott WO, Leng ER. Modern Corn Production. Champaign: A. & L. Publications; 1975

[2] Ranum. Global maize production, utilization, and consumption. Annals of the New York Academy of Sciences. 2014;**1312**(1):105-112

[3] Ahad MA. Trinojatiyo Fashaler Balai Bebosthapana (Pest Management in Gramicious Crops) (in Bengali). Dhaka, Bangladesh: Textbook Division, Bangla Academy; 2003. p. 184

[4] BBS. Statistical Pocket Book of Bangladesh. Dhaka: Bangladesh Bureau of Statistics, Planning Division, Ministry of Planning, Government of the Peoples Republic of Bangladesh; 2019

[5] BBS. Statistical Pocket Book of Bangladesh. Dhaka: Bangladesh Bureau of Statistics, Planning Division, Ministry of Planning, Government of the Peoples Republic of Bangladesh; 2020

[6] Covarelli L, Beccari G, Salvi S. Infection by mycotoxigenic fungalspecies and mycotoxin contamination of maize grain in Umbria, centralItaly. Food and Chemical Toxicology. 2011;**49**:2365-2369

[7] Magan N, Aldred D. Post-harvest control strategies: Minimizing mycotoxins in the field chain. International Journal of Food Microbiology. 2007;**119**(1-2):131-139

[8] Tsedaley B, Adugna G. Detection of fungi infecting maize (*Zea mays* L.) seeds in different storages around Jimma, Southwestern Ethiopia. Journal of Plant Pathology and Microbiology. 2016;7:3

[9] Richard JL, Payne GA. Mycotoxins: Risks in Plant, Animal and Human Systems. Ames, IA, USA: Council for Agricultural Science and Technology (CAST); 2003

[10] International Agency for Research on Cancer. IARC Monographs on the Evaluation of Carcinogenic Risks to Humans. Lyon, France: International Agency for Research on Cancer; 1993

[11] Food and Agriculture Organization. Worldwide regulations for mycotoxins in food and feed in 2003. In: FAO Food and Nutrition Paper no. 81. Rome: Food and Agriculture Organization; 2004. pp. 1-180

[12] Chu FS. Trichothecene mycotoxicoses. Encyclopedia. Human Biology. 1997;**8**: 511-522

[13] Reddy SV, Mayi DK, Reddy MU, Thirumala Devi K, Reddy DV. Aflatoxins B1 in different grades of chillies (*Capsicum annuum* L.) in India as determined by indirect competitive ELISA. Food Additives and Contaminants. 2001;**18**:553-558

[14] International Agency for Research on Cancer (IARC). Some traditional herbal medicines, somemycotoxins, naphthalene and styrene. IARC Monographs on the Evaluation of Carcinogenic Risks to Humans. 2002;**82**:1-556

[15] Mehl HL, Jaime R, Callicott KA, Probst C, Garber NP, Ortega-Beltran A, et al. *Aspergillus flavus* diversity on crops and in the environment can be exploited to reduce aflatoxin exposure and improve health. Belgrade-Zemun, Republic of Serbia: Annals of the New York Academy of Sciences, Institute for Animal Husbandry, Belgrade-Zemun, Republic of Serbia; 2012:1273-1277

[16] Jaime-Garcia R, Cotty PJ. Crop rotation and soil temperature influence the community structure of *Aspergillus flavus* in soil. Soil Biology and Biochemistry. 2010;**42**:1842-1847

[17] Probst C, Cllicott KA, Mehl HL, Jaime R. *Aspergillus flavus* diversity on crops and in the environment can be

exploited to reduce aflatoxin exposure and improve health. Annals of the New York Academy of Sciences. 2010;**1273**(1):7-17

[18] Agbetiameh D, Ortega-Beltran A, Awuah RT, Atehnkeng J, Cotty PJ, Bandyopadhyay R. Prevalence of aflatoxin contamination in maize and groundnut in Ghana: Population structure, distribution, and toxigenicity of the causal agents. Plant Disease. 2018;**102**(2018):764-772

[19] Scott PM, Trucksess WM. Mycotoxins in botanicals and dried fruits: A review, Food Additives and Contaminants—Part A Chemistry, Analysis, Control, Exposure and Risk Assessment. 2008;**25**(2):181-192

[20] Park KY, Bullerman LB. Effect of cycling temperatures on aflatoxin production by *Aspergillus parasiticus* and *Aspergillus flavus* in rice and Cheddar cheese. Journal of Food Science. 1983;**48**:889-896

[21] Abbas HK, Abbas MA, Locke RM, et al. Zablotowicz spatial variability of *Aspergillus flavus* soil populations under different crops and corn grain colonization and aflatoxins. Canadian Journal of Botany. 2004;**82**:1768-1775

[22] Lakkireddy K, Lakkireddy K, Kondapalli KS, et al. Aflatoxin in food and feed: The science of safe food. Research & Reviews: Journal of Food Science and Technology. 2014, 2014;**3**:6-11

[23] Bhuiyan MNH, Hassan MT, Begum M, Ahsan M, Rahim M. Occurrence and seasonal trends of aflatoxin in rice, maize and wheat in Bangladesh. IJSAT. 2013;**9**(8):1815-1272

[24] Fakruddin M, Chowdhury A, Hossain NM, Ahmed MM. Characterization of aflatoxin producing *Aspergillus flavus* from food and feed samples. SpringerPlus. 2015;**4**:159. DOI: 10.1186/s40064-015-0947

[25] Garcia R, Cotty PJ. *Aspergillus flavus* in soils and corncobs in South Texas: Implications for management of aflatoxins in corn-cotton rotations. Plant Disease. 2004;**88**:1366-1371

[26] Degola F, Berni E, Dall'Asta C, Spotti E, Marchelli R, Ferrero I, et al. A multiplex RT-PCR approach to detect aflatoxigenic strains of *Aspergillus flavus*. Journal of Applied Microbiology. 2007;**103**:409-417. DOI: 10.1111/j.1365-2672.2006.03256.x

[27] Ali N, Hossain K, Blaszkewicz M, Rahman M, Mohanto NC, Alim A, et al. Occurrence of aflatoxin M1 in urine from rural and urban adult cohorts in Bangladesh. Archives of Toxicology. 2016;**90**:1749-1755

[28] Dawlatana M, Coker RD, Nagler MJ, Wild CP, Hassan MS, Blunden G. The occurrence of mycotoxins in key commodities in Bangladesh: Surveillance results from 1993 to 1995. Journal of Natural Toxins. 2002;**11**:379-386

[29] Roy M, Harris J, Afreen S, Deak E, Gade L, Balajee SA, et al. Aflatoxin contamination in food commodities in Bangladesh. Food Additives & Contaminants. Part B, Surveillance. 2013;**6**:17-23

[30] Shephard GS, Westhuizen LV, Sewram V. Biomarkers of exposure to fumonisin mycotoxins: A review. Food Additives & Contaminants. 2003; **24**(10):1196-1201

[31] Glenn KC. Nutritional and safety assessments of foods and feeds nutritionally improved through biotechnology: Lysine maize as a case study. Journal of AOAC International. 2007;**90**:1470-1147

[32] Munkvold GP, Desjardins AE. Fumonisins in maize. Can we reduce their occurrence? Plant Disease. 1997;**81**:556-564

[33] Sutton JC. Epidemiology of wheat head blight and maize ear rot caused by *Fusariurn grcimineurum*. Ctmcidirrn Journd of Plant Pathology. 1982; **4**:195-209

[34] Rheeder JP, Marasas WF, Vismer H. Production of fumonisin analogs by Fusarium species. Applied and Environmental Microbiology. 2002; **68**(5):2101-2105

[35] Jens FC, Jørn S, Robert SA, Thomas LO, Ulf T. Fumonisin B2 production by *Aspergillus niger*. Journal of Agricultural and Food Chemistry. 2007;**55**(23):9727-9732

[36] Duan C, Qin Z, Yang Z, Li W. Identification of pathogenic Fusarium spp. causing maize ear rot and poten tial mycotoxin production in China. Toxins. 2016;**8**(6):186

[37] Alexander NJ, Proctor RH, McCormick SP. Genes, gene clusters, and biosynthesis of trichothecenes and fumonisins in *Fusarium*. Toxin Reviews. 2009;**28**:198-215. DOI: 10.1080/15569540903092142

[38] Baldwin T, Riley R, Zitomer N, Voss K, Coulombe R Jr, Pestka J, et al. The current state of mycotoxin biomarker development in humans and animals andthe potential for application to plant systems. World Mycotoxin Journal. 2011;**4**(3):257-270

[39] Schmidt-Heydt M, Geisen R. Gene expression as an indication for ochratoxin A biosynthesis in *Penicillium nordicum*. Mycotoxin Research. 2007; **23**:13-21

[40] Schmidt-Heydt M, Magan N, Geisen R. Stress induction of mycotoxin biosynthesis genes by abiotic factors. FEMS Microbiology Letters. 2008;**284**: 142-149

[41] De Boevre M, Mavunguose DJ, Landschoot S, Audenaer K. Natural occurrence of mycotoxins and their masked forms in food and feed products. World Mycotoxin Journal. 2012;**5**(3):207-219

[42] Fanelli F, Iversen A, Logrieco A, Mule G. Relationship between fumonisin production and FUM gene expression in *Fusarium* verticillioides under different environmental conditions, Food Additives and Contaminants—Part A Chemistry, Analysis, Control, Exposure and Risk Assessment. 2012;**30**(2):365-371

[43] Proctor RH, Van Hove F, Susca A, Stea G, Busman M, Van der Lee T, et al. Birth, death and horizontal gene transfer of the fumonisin byosinthetic gene cluster during the evolutionary diversification of *Fusarium*. Molecular Microbiology. 2013;**90**:290-306. DOI: 10.1111/mmi.12362

[44] Desjardins AE. *Fusarium* Mycotoxins. Chemistry, Genetics and Biology. St Paul, MN: APS Press; 2006

[45] Battilani P, Camardo LM, Rossi V, Giorni P. AFLA-maize, a mechanistic model for *Aspergillus flavus* infection and aflatoxin B1 contamination in maize. Computers and Electronics in Agriculture. 2013;**94**:38-46

[46] EFSA. Scientific opinion on the risks for public health related to the presence of zearalenone in food. EFSA Journal. 2011;**9**(6):2197

[47] Jindal N, Mahipal SK, Rottinghaus GE. Occurrence of fumonisin B_1 in maize and poultry feeds in Haryana, India. Mycopathologia. 1999;**148**:37-40

[48] Jakic-Dimic D, Nesic K. Mycotoxins in feed. In: Proceedings of the XIII Symposium Feed Technology. Novi Sad; 2009;**1273**(1):7-17

[49] Gruber-Dorninger C, Jenkins T, Schatzmayr G. Global mycotoxin

occurrence in feed: A ten-year survey. Toxins (Basel). 2019;**11**(7):375

[50] Phillips SI, Wareing PW, Dutta A, Panigrahi S, Medlock V. The mycoflora and incidence of aflatoxin, zearalenone and sterigmatocystin in dairy feed and forage samples from Eastern India and Bangladesh. Mycopathologia. 1996; **133**(1):15-21

[51] Giasuddin M, Sil BK, Alam J, Koike I, Islam MR, Rahman MM. Prevalence ofpoultry diseases in Bangladesh. Journal of Biological Sciences. 2002;**2**:212-213

[52] Abdel-Wahhab MA, Abdel-Galil MM, Hassan AM, Hassan NH, Nada SA, Saeed A, et al. Zizyphus spina-christi extract protects against aflatoxin B1-intitiated hepatic carcinogenicity. African Journal of Traditional, Complementary, and Alternative Medicines. 2007;**4**:248-256

[53] Tejada AW, Rustia A. Risk assessment of contaminants in foods: Mycotoxins and pesticide residues. In: Proceedings of FFTCeKU 2011 Conference, International Seminar on Risk Assessment and Risk Management of Mycotoxins for Food Safety in Asia. Thailand: Kasetsart University; 2011

[54] Yoshizawa T. A current situation of mycotoxin management in Asia in relation to recent actions in Japan. In: Proceedings of FFTCeKU 2011 Conference, International Seminar on Risk Assessment and Risk Management of Mycotoxins for Food Safety in Asia. Thailand: Kasetsart University; 2011

[55] Mahfuz M, Alam MA, Fahim SM, Jyoti MR, Hossain M, Egner PA, et al. Aflatoxin exposure in children living in Mirpur, Dhaka: Data from MAL-ED companion study. Journal of Exposure Science & Environmental Epidemiology. 2019;**29**:655-662

[56] Van Egmond HP, Schothorst RC, Jonker MA. Regulations relating to mycotoxins in food. Analytical and Bioanalytical Chemistry. 2007;**389**: 147-157

[57] Moss MO. Mycotoxin review—2 Fusarium. Mycologist. 2002;**16**:158-161

[58] Ahlberg SH, Joutsjoki V, Korhonen HJ. Potential of lactic acid bacteria in aflatoxin risk mitigation. International Journal of Food Microbiology. 2015;**207**:87-102

[59] Stevens DB, Turner JA, Paveley ND. Exploiting variety resistance to rationalise fungicide inputs—Theory and practice. Aspects of Applied Biology. 1997;**50**:279-284

[60] Missmer SA, Suarez L, Felkner M, Wang E, Merrill AH Jr, Rothman KJ, et al. Exposure to fumonisins and the occurrence of neural tube defects a long the Texas-Mexico border. Environmental Health Perspectives. 2006;**114**:237-241

[61] Covarelli L, Stifano S, Beccari G, Raggi L, Lattanzio VMT, Albertini E. Characterization of Fusarium verticillioides strains isolated from maize in Italy: Fumonisin production, pathogenicity and genetic variability. Food Microbiology. 2012;**31**:17-24

[62] Gelderblom WCA, Jaskiewicz K, Marasas WFO, Thiel PG, Horak RM, Vleggaar R, et al. Fumonisins-Novel mycotoxins with cancer-promoting activity produced by *Fusarium moniliforme*. Applied and Environmental Microbiology. 1988;**54**:1806-1811

[63] Turna NS, Wu F. Risk assessment of aflatoxin-related liver cancer in Bangladesh. Food Additives & Contaminants: Part A: Chemistry, Analysis, Control, Exposure & Risk Assessment. 2019;**36**(2):320-326

[64] ISTA. International Rules for Seed Testing. Bassersdorf, Switzerland: International Seed Testing Association; 2006

[65] Sreenu B, Girish A, Alice J, Sujeetha R. Identification and detection of maize seed borne pathogens using different seed testing methods. International Journal of Current Microbiology and Applied Sciences. 2019;**8**:1460-1466

[66] Dinu D, Nechifor MT, Stoian G, Costache M, Dinischiotu A. Enzymes with new biochemical properties in the pectinolytic complex produced by *Aspergillus niger* MIUG 16. Journal of Biotechnology. 2007;**131**:128-137

[67] Mlakar T, Legiša M. Citrate inhibition-resistant form of 6-phosphofructo-1-inase from *Aspergillus niger*. Applied and Environmental Microbiology. 2006;**72**:4515-4521

[68] Probst AV, Dunleavy E, Almouzni G. Epigenetic inheritance during the cell cycle. Nature Reviews. Molecular Cell Biology. 2009;**10**:192-206

[69] Navi SS, Bandyopadhyay R, Hall AJ, Bramel-Cox PJ. A pictorial guide for the identification of mold fungi on sorghum grain. In: Information Bulletin no. 59. Patancheru 502 324, Andhra Pradesh, India: International Crops Research Institute for the Semi-Arid Tropics; 1999. 118 pp

[70] White TJ, Bruns TD, Lee S, Taylor J. Amplification and direct sequencing of fungal ribosomal genes form phylogenetics. In: Innis MA, Gelfrand DH, Sninsky JJ, White TJ, editors. PCR Protocols. San Diego, California, USA: Academic Press; 1990. pp. 315-322

[71] Criseo G, Bagnara A, Bisignano G. Differentiation of aflatoxin-producing and non-producing strains of *Aspergillus flavus* group. Letters in Applied Microbiology. 2001;**33**:291-295

[72] Bluhm BM, Cousin MA, Woloshuk CP. Multiplex real-time PCR detection of fumonisin-producing and trichothecene- producing groups of Fusarium species. Journal of Food Protection. 2004;**3**:536-543

[73] Shapira R, Paster N, Eyal O, Menasherov M, Mett A, Salomon R. Detection of aflatoxigenic molds in grains by PCR. Applied and Environmental Microbiology. 1996;**62**: 3270-3273

[74] Chen RS, Tsay JG, Huang YF, Chiou RYY. Polymerase chain reaction mediated characterization of molds belonging to the *Aspergillus flavus* group and detection of *A. parasiticus* in peanut kernels by multiplex polymerase chain reaction. Journal of Food Protection. 2002;**65**:840-844

[75] Mayer Z, Bagnara A, Färber P, Geisen R. Quantification of the copy number of nor-1, a gene of the aflatoxin biosynthetic pathway by real-time PCR, and its correlation to the cfu of *Aspergillus flavus* in foods. International Journal of Food Microbiology. 2003;**82**:143

[76] Ibrahim F, Jalal H, Khan AB, Asghar MA, Iqbal J, Ahmed A, et al. Prevalence of aflatoxigenic Aspergillus in food and feed samples from Karachi, Pakistan. JIMB. 2016;**4**(1):1-8

[77] Levin ER. PCR detection of aflatoxin producing fungi and its limitations. International Journal of Food Microbiology. 2012;**156**:1-6

[78] Ana A, Savoie J-M, Chereau S, Ducos C, Aguilar M, et al. Primingto protect maize from *Fusarium verticillioides* and its fumonisin accumulation. Journal of the Science of Food and Agriculture. 2019;**99**(1):64-72

[79] Monika R, Julie H, Sadia A, Eszter D, Lalitha G, S, A. B., & Stephen, L. Aflatoxin contamination in food commodities in Bangladesh. Food Additives & Contaminants: Part B: Surveillance. 2013;**6**(1):17-23. DOI: 10.1080/19393210.2012.720617

[80] Rheeder JP, Marasas WFO, van Schalkwyk DJ. Incidence of *Fusarium* and *Diploidia* species in naturally infected grain of South African maize cultivars: A follow-up study. Phytophylactica. 1993;**25**:43-48

[81] Waliyar F, Reddy SV, Lava-Kumar P. Review of immunological methods for the quantification of aflatoxins in peanut and other foods. Peanut Science. 2009;**36**:54-59. DOI: 10.3146/AT07-007.1

[82] Ketney O, Santini A, Oancea E. Recent Aflatoxin survey data in milk and milk products: A review. International Journal of Dairy Technology. 2017;**70**:320-331. DOI: 10.1111/1471-0307.12382

[83] Andreasson U, Perret-Liaudet A, Waalvijk van Doorn LJC, Perret-Liaduet A, Blennov K, Chiasserini D, et al. A practical guide to immunoassay method validation. Frontiers in Neurology. 2015;**6**:179

[84] Czéh Á, Mandy F, Feher-Toth S, Torok L, Mike Z, Koszegi B, et al. A flow cytometry based competitive fluorescent microsphere immunoassay (CFIA) system for detecting up to six mycotoxins. Journal of Immunological Methods. 2012;**384**:71-80. DOI: 10.1016/j.jim.2012.07.010

[85] Czéh Á. Mikrogyöngy Alapú Multiplex Immunoassay Rendszer Fejlesztése Multi-mikotoxin Vizsgálatra: Túl a XX. Századi Alkalmazásokon [master's thesis, Interdiszciplináris Orvostudományok Doktori Iskola D93]. Pécs: University of Pécs; 2014

[86] Bánáti H, Darvas B, Fehér-Tóth S, Czéh Á, Székács A. Determination of mycotoxin production of *Fusarium* species in genetically modified maize varieties by quantitative flow immunocytometry. Toxins. 2017;**9**:70. DOI: 10.3390/toxins9020070

[87] Kumar Ajith K, Naik MK. Prevalence and distribution of aflatoxin contamination of chilli (*Capsicum annuum* L.) field and market. Karnataka. The Journal of Agricultural Science. 2005;**18**(2):520-523

[88] Lopez-Errasquin E, Vazquez C, Jimenez M, Gonzalez-Jaen MT. Real-time RT-PCR assay to quantify the expression of *fum1* and *fum19* genes from the fumonisin-producing *Fusarium verticillioides*. Journal of Microbiological Methods. 2007;**68**:312-317

[89] Williams WP. Breeding for resistance to aflatoxin accumulation in maize. Mycotoxin Research. 2006;**22**: 27-32

[90] Saito M, Machida S. A rapid identification method for aflatoxin producing strains of *A. flavus* and *A. parasiticus* by ammonia vapor. Mycoscience. 1999;**40**:205-222

[91] Bennett JW, Lee LS. Mycotoxins—Their biosynthesis in fungi: Aflatoxins and other bisfuranoids. Journal of Food Protection. 1979;**42**:805-809

[92] Gary M. Epidemiology of *Fusarium* diseases and their mycotoxins in maize ears. European Journal of Plant Pathology. 2003, 2003;**109**(7):705-713

[93] Maiorano A, Reyneri A, Magni A, Ramponi C. A decision tool for evaluating the agronomic risk of exposure to fumonisins of different maize crop management systems in Italy. Agricultural Systems. 2009;**102**: 17-23

[94] Cao A, Santiago R, Ramos AJ, Souto XC, Aguín O, Malvar RA, et al. Critical environmental and genotypic factors for *Fusarium verticillioides* infection, fungal growth and fumonisin contamination in maize grown in northwestern Spain. International Journal of Food Microbiology. 2014;**177**:63-71. DOI: 10.1016/j.ijfoodmicro.2014.02.004

[95] Czembor E, Stepien Ł, Waśkiewicz A. Effect of environmental

factors on *Fusarium* species and associated mycotoxins in maize grain grown in Poland. PLoS One. 2015;**10**: e0133644. DOI: 10.1371/journal.pone.0133644

[96] Cendoya E, Chiotta ML, Zachetti V, Chulze SN, Ramirez ML. Fumonisins and fumonisin-producing Fusarium occurrence in wheat and wheat by products: A review. Journal of Cereal Science. 2018;**80**:158-166

[97] Chen ZY, Brown RL, Cleveland TE. Evidence for an asso- ciation in corn between stress tolerance and resistance to *Aspergillus flavus* infection and aflatoxin contamination. African Journal of Biotechnology. 2004;**3**:693-699

[98] Magan N, Aldred D. Why do fungi produce mycotoxins? In: Dijksterhuis J, Samson R, editors. Food Mycology: A Multifaceted Approach to Fungi and Food. Boca Raton, FL, USA: CRC Press; 2007

[99] Jurado M, Vázquez C, Callejas C, González-Jaén M. Occurrence and variability of mycotoxigenic Fusarium species associated to wheat and maize in the South West of Spain. Mycotoxin Research. 2006;**22**:87-91

[100] Aliakbari F, Mirabolfathy M, Emami M, Mazhar SF, Karami-Osboo R. Natural occurrence of Fusarium species in maize kernels at Gholestan province in northern Iran. Asian Journal of Plant Sciences. 2007;**8**:1276-1281

[101] Cavaglieri L, Keller K, Pereyra C, Pereyra MG, Alonso V, Rojo F, et al. Fungi and natural incidence of selected mycotoxins in barley rootlets. Journal of Stored Products Research. 2009;**45**: 147-150

[102] Gil-Serna J, Mateo E, González-Jaén M, Jiménez M, Vázquez C, Patiño B. Contamination of barley seeds with *Fusarium* species and their toxins in Spain: An integrated approach. Food Additives & Contaminants: Part A. 2013;**30**:372-380

[103] Magan N, Medina A, Aldred D. Possible climate-change effects on mycotoxin contamination of food crops pre-and postharvest. Plant Pathology. 2011;**60**:150-163

[104] Marasas FO, W. Discovery and occurrence of the fumonisins: A historical perspective. Environmental Health Perspectives. 2001;**109**(suppl. 2): 239-243

[105] Abbas HK, Zablotowicz RM, Bruns HA. Modeling of colonization of maize by toxigenic and non-toxigenic *Aspergillus flavus* strains: Implication for biological control. World Mycotoxin Journal. 2008;**1**:333-340

[106] Jian F, Jayas DS. The ecosystem approach to grain storage. Agricultural Research. 2012;**1**(2):148-156

[107] Johnson LA. Corn production, processing and utilization. In: Lorenz KJ, Kulp K, editors. Handbook of Cereal Science and Technology. Vol. 1. New York, U.S.A: Marcel Dekker Inc; 1991. pp. 55-131

[108] Magan N, Hope R, Cairns V, Aldred D. Post-harvest fungal ecology: Impact of fungal growth and mycotoxin accumulation in stored grain. European Journal of Plant Pathology. 2003;**109**(7): 723-730

[109] Jayas DS, White Noel DG. Storage and drying of grain in Canada: Low cost approaches. Food Control. 2003;**14**(4): 255-261

[110] Placinta C, D'Mello JP, Macdonald AM. A review of worldwide contamination of cereal grains and animal feed with *Fusarium* mycotoxins. Animal Feed Science and Technology. 1999;**78**:21-37

[111] Bennett JW, Klich M. Mycotoxins. Clinical Microbiology Reviews. 2003;**16**:497-516

[112] Richard JL. Some major mycotoxins and their mycotoxicoses—An overview. International Journal of Food Microbiology. 2007;**119**:3-10

[113] Streit E, Schatzmayr G, Tassis P, Tzika E, Marin D, Taranu I, et al. Current situation of mycotoxin contamination and co-occurrence in animal feed—Focus on Europe. Toxins. 2012;**4**:788-809

[114] Bryden WL. Mycotoxin contamination of the feed supply chain: Implications for animal productivity and feed security. Animal Feed Science and Technology. 2012;**173**:134-158

[115] Arunachalam C, Doohan FM. Trichothecene toxicity in eukaryotes: Cellular and molecular mechanisms in plants and animals. Toxicology Letters. 2013;**217**:149-158

[116] Abbas HK, Yoshizawa T, Shier WT. Cytoxicity and phytotoxicity of trichothecene mycotoxins produced by *Fusarium* spp. Toxicon. 2013;**74**:68-75

[117] Bhattacharjee R, Dey U. An overview of fungal and bacterial biopesticides to control plant pathogens/diseases. African Journal of Microbiology Research. 2014;**8**: 1749-1762

[118] Mukhopadhyay R, Kumar D. Trichoderma: A beneficial antifungal agent and insights into its mechanism of biocontrol potential. Egyptian Journal of Biological Pest Control. 2020;**30**:133

[119] Dorner JW, Cole RJ. Effect of application of nontoxigenic strains of *Aspergillus flavus* and *A. parasiticus* on subsequent aflatoxin contamination of peanuts in storage. Journal of Stored Products Research. 2002;**38**:329-339

Chapter 6

Critical Dry Spell Prediction in Rain-Fed Maize Crop Using Artificial Neural Network in Nigeria

Nnadozie Okonkwo Nnoli, Ahmed Balogun, Jerome Omotosho and Samuel Agele

Abstract

Prediction of yearly mid-growing season first and second critical dry spells using artificial neural networks (ANN) for enhanced maize yield in nine stations in Nigeria is performed. The ANN model uses nine meteorological parameters to predict onset dates and lengths of the critical dry spells. The daily dataset is from 1971 to 2013 of which about 70% is used for training while 30% is for testing. Seven ANN models are developed for each station with a view to measuring their predictive ability by comparing predicted values with the observed ones. Prediction lead times for the two critical dry spell onset dates generally range from about 2 weeks to 2 months for the nine stations. Error range during testing for the onset dates and lengths of first and second critical dry spells is generally ±4 days. The root-mean-square error (RMSE), coefficient of determination, Nash-Sutcliffe coefficient of efficiency, Wilmott's index of agreement, and RMSE observation standard deviation ratio range from 0.46 to 3.31, 0.58 to 0.93, 0.51 to 0.90, 0.82 to 0.95, and 0.30 to 0.69, respectively. These results show ANN capability of making the above reliable predictions for yearly supplementary irrigation planning, scheduling, and various other decision makings related to sustainable agricultural operations for improved rain-fed maize crop yield in Nigeria.

Keywords: Nigeria, rain-fed maize, critical dry spells, yearly prediction, artificial neural network

1. Introduction

Variability of rainfall in Nigeria as well as in West Africa, etc., leads to the occurrence of wet and dry spells within the growing season. Short- and long-duration dry spells are noted during the period of crop growth and development on yearly basis. Song et al. [1] using weather- and county-level maize yield data estimated the drought risk for maize in China for the period from 1971 to 2010. They noted that drought risk had increased in China over the last 40 years and that the reasons for the observed changes were increased drought hazard associated with climate change and increased exposure of maize to drought due to an expanded production area. Significant drought incidents have seriously affected sustainable agriculture,

IntechOpen

people's living condition, and the economy of many developing and under-developed countries [2, 3]. The occurrence and distribution of dry spells, especially the longer ones at critical times during growing season, generally have negative impact on maize crop development and yield under rain-fed farming in Nigeria. According to Mugalavai et al. [4] and Gao et al. [5], the most critical growth stages for maize crop in terms of dry spell occurrences are the germination, tasseling, and flowering. Germination is within the initial stage, while tasseling and flowering occur during the mid-season stage of growing season. The four crop growth stages are initial, development, mid-season, and late season [6]. Advance knowledge on critical dry spell onset dates and lengths for rain-fed maize crop on yearly basis is very important in supplementary irrigation planning, scheduling, and various other decision makings related to sustainable agricultural operations for improved maize yield.

Sharma [7] noted that a major challenge of drought research was to develop suitable methods and techniques for forecasting the onset and termination points of droughts. Successful development of suitable methods will enable stakeholders in agricultural and water resource sectors of the economy to embark upon risk-based (proactive) rather than crisis-based (reactive) approach to drought management in areas prone to drought [8, 9]. This is also applicable to dry spell management. Most publications are concerned with probabilistic, statistical, and stochastic modeling, and the most widely used stochastic models are autoregressive integrated moving average (ARIMA) models [10]. A dynamical model and a statistical model have been used to determine trends and make seasonal predictions of rainfall and dry spells occurrence in Ghana [11].

In recent years, neural-based models have been gaining attention over statistical models, possibly owing to the simplicity in modeling complex problems when many parameters are taken into consideration [12]. Abrishami et al. [13] used artificial neural network (ANN) model for estimating wheat and maize daily standard evapotranspiration. The results showed the suitable capability and acceptable accuracy of ANN. Mulualem and Liou [14] developed seven ANN predictive models incorporating hydro-meteorological, climate, sea surface temperatures, and topographic attributes to forecast the standardized precipitation evapotranspiration index (SPEI) for seven stations in the Upper Blue Nile basin (UBN) of Ethiopia from 1986 to 2015. Statistical comparisons of the different models showed that accurate results in predicting SPEI values could be achieved by including large-scale climate indices. Morid et al. [15] were able to show the efficiency of ANN when it was used for forecasting some drought indices in some selected places in Iran for up to 12 months lead times [3]. One neural network model was developed to forecast precipitation occurrences such as "rain" or "no-rain," while another model was developed to predict the amount of precipitation at several sub-levels using fuzzy techniques in Sri Lanka [16]. The ability of neural network model to predict "no-rain" situation gives it credence to forecast dry spell. Mathugama and Peiris [17] therefore recommended the exploration of the use of artificial neural network (ANN) to predict dry spell properties and that the models had to be statistically validated. Studies related to forecasting critical dry spell onset dates and lengths (especially mid-growing season dry spells) in Nigeria and other places are scarce. Farmers (especially maize farmers) in Nigeria desire to know on yearly basis when dry spells—critical dry spells—will occur after planting their crops to enable them plan their yearly agricultural operations effectively. In Nigerian Meteorological Agency (NiMet), numerical model have been used for sub-seasonal to seasonal forecasts of weather elements [18], while statistical models are used in seasonal rainfall forecasts for agricultural activities. Probabilistic forecasts have been made [19] for severe dry spell occurrences of lengths 10–21 days and moderate ones of lengths 8–15 days for 10 northern States for the month of June for year 2020; however, specific dry spells onset dates are not given.

These informed our embarking on this study in aid of effective yearly agricultural operations for improved crop yield and maize in particular. The objective therefore of the present work is to predict the onset dates and lengths of mid-growing season critical dry spells for rain-fed maize crop *on yearly basis* in Nigeria using artificial neural network (ANN) model to enable farmers in those stations plan *yearly* agricultural operations for enhanced maize yield.

2. Study area, data, and methodology

2.1 Study area

Table 1 shows the geographical and some climate characteristics of the study area. The following nine meteorological stations in their respective agro-ecological zones in Nigeria are considered: Calabar, Warri, Ibadan, Makurdi, Lokoja, Ilorin, Yola, Kaduna, and Yelwa.

2.2 Data

a. The data used for this work are as follows: the daily maximum, minimum, and mean temperatures, 0600 and 1500 GMT relative humidity, wind speed at 2 meter level, and sunshine hours (1971–2013) for the nine stations from Nigerian Meteorological Agency (NiMet), Oshodi, Lagos and supplemented with 0.125° resolution ERA INTERIM Reanalysis data (1979–2013) [20].

b. NiMet daily rainfall data supplemented with the daily 0.25° horizontal resolution 3B42 rainfall from Tropical Rainfall Measuring Mission (1998–2013) [21].

Since the NiMet data were supplemented as stated above, the Adapted Caussinus-Mestre Algorithm for homogenizing Networks of Temperature series (ACMANT) was used to check and correct the inhomogeneities in the quality controlled time series. A full scientific description of ACMANT setup could be found in [22]. Several studies included [22, 23], etc. have been effectively used ACMANT in homogenizing series. Good performances of homogenizing climatic series with ACMANT are noted in the result evaluation from these studies.

Agro-ecological zone	Station	Long.	Lat.	Elev. (m)	Annual rainfall (mm/year) (1971–2013)		
					Max.	Min.	Mean
Northern Guinea Savannah	Yelwa	4.75°E	10.88°N	244.0	1564.6	388.9	986.5
	Kaduna	7.45°E	10.60°N	641.0	1659.8	793.4	1211.1
	Yola	12.47°E	9.23°N	190.5	1142.7	468.5	873.4
Southern Guinea Savannah	Ilorin	4.58°E	8.48°N	344.0	1539.3	697.7	1177.6
	Lokoja	6.73°E	7.80°N	62.5	1767.1	771.7	1196.4
	Makurdi	8.53°E	7.75°N	91.4	1617.1	761.5	1182.8
Rain Forest	Ibadan	3.90°E	7.43°N	220.7	1967.8	775.7	1328.9
Mangrove Swamp	Warri	5.73°E	5.52°N	6.0	3414.4	2051.5	2734.3
	Calabar	8.33°E	4.95°N	62.3	4044.9	2109.5	2937.6

Table 1.
Area of study showing the stations, agro-ecological zones and climate characteristics.

2.3 Methodology

2.3.1 Growing season onset and cessation dates

The determination of onset and cessation dates of growing season was carried out using the methods of [24, 25]. The onset date of growing season was defined by [24] for northern Nigeria as the date when accumulated daily rainfall exceeded 0.5 of the accumulated reference evapotranspiration for the remainder of the season, provided that no dry spell of 5 days or more occurred in the week after that date. The determination of onset date of rains according to [25] was from the first point of maximum positive curvature of the plotted graph of cumulative percentage of pentade rainfall, while cessation was from the last point of maximum negative curvature of the plotted graph of cumulative percentage of pentade rainfall. The method of [25] was initially used to determine the onset dates of growing season, while that of [24] was next used to ensure that no dry spell of 5 days or more occurred in the week after that date.

2.3.2 Reference evapotranspiration

To determine the critical dry spells during each growing season, the daily reference evapotranspiration (ET_o) was first computed using the FAO Penman-Monteith Equation [6]. This equation, given below (Eq. (1)), used the abovementioned data with the exception of daily rainfall.

$$ET_o = \frac{0.408\Delta(R_n - G) + \gamma\left(\frac{900}{T+273}\right)u_2(e_s - e_a)}{\Delta + \gamma(1 + 0.34u_2)} \quad (1)$$

where T—air temperature at 2 m height (°C), u_2—wind speed at 2 m height (ms^{-1}), e_s—saturation vapour pressure (kPa), e_a—actual vapour pressure (kPa), ($e_s - e_a$)—saturation vapour deficit (kPa), R_n—net radiation at the crop surface (MJm^{-2} day^{-1}), G—soil heat flux density (MJm^{-2} day^{-1}), Δ—slope vapour pressure curve (kPa °C^{-1}), γ—psychrometric constant (kPa °C^{-1}).

The above equation determines the evapotranspiration (ET_o) from the hypothetical grass reference surface. The effect of soil heat flux (G) is ignored for daily calculations [6] as the magnitude of the heat flux in this case is relatively small. The FAO Penman-Monteith method [6] is still used as the sole standard method as could be seen in recent research work on reference evapotranspiration included in [26, 27], etc. However, since the number of requested climatic variables is often not available under limited data conditions [28, 29], other simple ETo equations with less number of requested climatic variables have been used to compute ETo values that are close to the FAO Penman-Monteith method. These methods are the four of the Valiantzas equations, along with the Makkink, Calibrated Hargreaves, Abtew, Jensen-Haise, and Caprio equations and could be used as best alternative ET_o estimation methods. These alternative equations could be used across the dry semi-arid and arid zones where water is the most limiting factor to food and fiber production [27].

The maize crop variety used in this study is the 118-day one whose phenology is as follows: 20 days for initial, 32 days for development, 38 days for mid-season, and 28 days for late season growth stages. This is based on the what is stated in [6] that the length of crop development stages provided in their table is indicative of general conditions; the user is therefore strongly encouraged to obtain appropriate local information.

2.3.3 Dry day and critical dry spell definition

It is usual to use rainfall thresholds higher than zero millimeter to define a dry day in order to account for the measurement errors or very little amounts of rain that are not available for plants or water resources, due to interception and/or direct evaporation [29, 30]. Different precipitation thresholds of 1–10 mm/day but fixed for the whole observation period are considered by most authors in analyzing long dry spells [30–33]. However, since the evaporation varies throughout the year and for different locations, fixed rainfall thresholds are not representative of real ground conditions. The net precipitation that is available for plants can be strongly modulated by atmospheric evaporative demands thereby affecting water stress levels by plants and crops [34–36]. Meteorological data from different approaches such as potential evaporation [37] or the reference evapotranspiration [6] can be used to determine atmospheric evaporative demand. Rivoire et al. [38] emphasized the need to take account of the atmospheric evaporative demand instead of making use of fixed rainfall thresholds for defining a dry day when analyzing dry spells with respect to agricultural impacts in particular. A dry day in this work is therefore taken as the day when the rainfall (RR) is less than the average reference evapotranspiration, ET_o [38, 39]. A threshold of ET_o is considered to define a dry day when $RR–ETo \leqq 0$ [38]. A number of these consecutive dry days constitute a spell. The critical dry spells are therefore those that occur at germination/emergence and establishment (initial stage), and close to and during the tasseling and flowering stage (mid-season stage). However, the critical dry spell prediction carried out in this work is for the mid-season stage only. **Figure 1** shows the time series of rainfall and mean reference evapotranspiration against day of the year from around planting date to harvesting date for maize crop for 1973 in Ibadan. Four critical dry spells during the mid-season are indicated. The minimum number of consecutive dry days that constitute a spell in this work is 3 days [40].

2.3.4 Artificial neural network (ANN) model

2.3.4.1 Model description

Artificial neural network (ANN) model was used in this work for the prediction of mid-growing season critical dry spell onset dates and lengths. ANN is a "black box" model of a type that is often used to model high-dimensional nonlinear data. It is a nonstatistical data modeling tool, which is contained in any version of R statistic or Matlab tool box. ANN is a highly interconnected network of machine learning algorithm based on the model of a human neuron. It mimics this model or structure by distributing its computations to small and simple processing units called artificial neurons or nodes [41, 42]. Artificial intelligence (AI) makes it possible for machines to learn from experience, adjust to new inputs, and perform human-like tasks. According to [42], ANN is data-driven, self-adaptive methods since there are few known assumptions about the models for problems under study unlike the traditional model-based methods. ANN model learns from examples and captures subtle functional relationships among the data even if the underlying relationships are unknown or hard to describe. This makes ANN very appropriate for problems whose solutions require knowledge that is not easy to state explicitly but for which there are enough observations [42]. Therefore, they can be treated as one of the multivariate nonlinear nonparametric statistical methods [43, 44]. After learning the data presented to them, ANN can generalize and often correctly infer the unseen part of a population even if the sample data contain noisy information. Since ANNs can compute the value of any continuous function to any desired accuracy as has been shown by [45–47], they are considered as universal functional approximators [42].

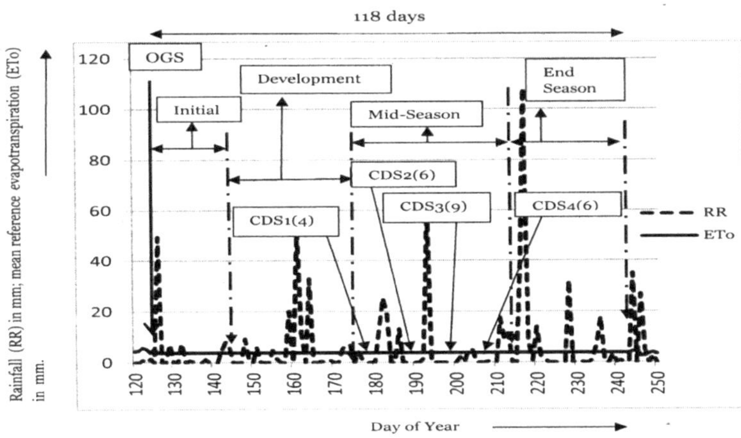

Figure 1.
Time series of rainfall (RR in mm) and mean reference evapotranspiration (ETo in mm) from the onset date of growing season, OGS (in days of year) for 118-day maize crop to its harvesting time for 1973 in Ibadan. $CDS_1(4)$, $CDS_2(6)$, $CDS_3(9)$, and $CDS_4(6)$ represent the first, second, third, and fourth critical dry spells with lengths 4, 6, 9, and 6 days in brackets, respectively. Mean (1971–2013) reference evapotranspiration for growing season for maize is approximately 3.72 mm.

ANN is made up of three layers of units, the input, hidden, and output layers. The ANN receives the input signal from the external world in the form of a pattern and image in the form of a vector. Each of the input is then multiplied by its corresponding weights. These weights are the details used by the ANNs to solve a certain problem. The activity of the input layer represents the raw information that is fed into the network, while the activities of each hidden layer are determined by the activities of the input layer and the weights on the connections between the input and the hidden layer. The behavior of the output layer depends on the activity of the hidden layer and the weights between the hidden and the output layers. To train the neural network models, the training parameters for the chosen algorithm must be specified in terms of the inputs, the number of hidden and output layer neurons, and the activation function of each layer [41, 48]. To fulfill these requirements, the correct number of regressor as well as the number of hidden neurons must first be selected but there are no specific rules for these selections [49, 50]. In many applications, the number of neurons for the hidden layer is selected based on trial-and-error method usually starting with small initial network [51]. A sample ANN architecture for first critical dry spell onset date prediction for Ibadan is shown in **Figure 2** having one input layer of nine neurons, two hidden layers—first of nine neurons and second of two neurons—and one output layer of one neuron. The inputs are multiplied by modifiable weights that are crucial parameters of the ANN models for solving a problem. ANN model could be run in R software—in R studio [52]. The neuralnet package (neuralnet) version 1.33 of August 5, 2016 [53] was used in this work. The training of neural networks uses the back-propagation, resilient back-propagation with [54] or without weight backtracking [55], or the modified globally convergent version by [56]. The package allows flexible settings through custom choice of error and activation function and it can combine fast convergence and stability and generally provides good results [55]. Furthermore, the calculation of generalized weights [58] is implemented. In this work, the default neural algorithm was used (i.e., "rprop+"). This refers to the resilient back-propagation with weight backtracking [54]. Amid the pool of the weight-updating process, the resilient back-propagation (RProp algorithm from the "nuerlanet" package in R) was chosen because it can combine fast convergence and stability and generally provides good results.

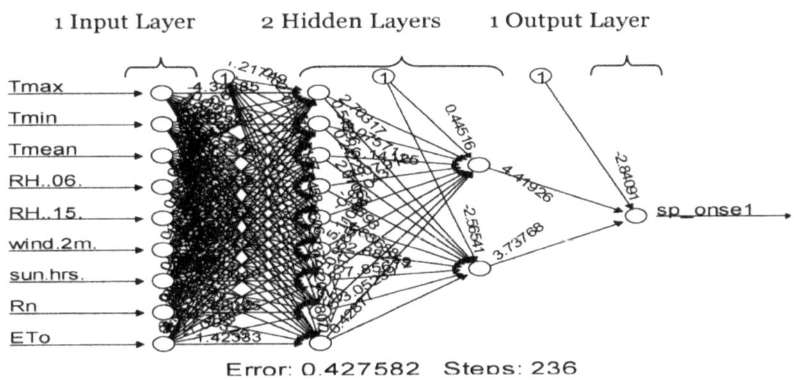

Figure 2.
A sample ANN architecture for first critical dry spell onset date prediction for 2008 for Ibadan, Nigeria.

2.3.4.2 Prediction procedure

The data attributes (predictors) used in the neural network model were maximum, minimum, and mean temperatures (T_{max}), (T_{min}), and (T_{mean}), relative humidity at 0600 (RH_{06}) and 1500 (RH_{15}) GMT, wind speed at 2 m level (u_2), sunshine hours (n), net radiation (R_n), and reference evapotranspiration (ET_o), while the data classes (predictands) were onset dates and lengths of critical dry spells. Net radiation and reference evapotranspiration were computed. In this work, the 43-year data were partitioned into two: 30 years for training and 13 independent years for testing. The test data were kept out of the process of producing the ANN model in order to test its predictive power [14]. This corresponds to approximately 70% (two-thirds) of data for training and 30% (one-third) for testing. Regarding the data partitioning, some authors have used two-thirds of data for training and one-third for validation and testing [57–59]. Dubey [58] used approximately half of one-third of data each for validation and testing. However, Mulualem and Liou [14] who worked on the application of artificial neural networks in forecasting a standardized precipitation evaporative index (SPEI) for the Upper Blue Nile Basin, Ethiopia (using RProp algorithm from the "neuralnet" package in R), partitioned their data into training and test sets. This method was applied in this work. The two independent datasets were chosen in such a way that early, normal, and late onset dates of critical dry spells were reflected in each of them. Likewise, the lowest, normal, and highest lengths of critical dry spells were also reflected in each of them. The neural network architecture consists of one input layer, one or two hidden layers, and one output layer. The input layer (first layer) of neurons consists of the nine attributes (predictors). The hidden layers of neurons are two (the second and third layers) and in few cases one. The hidden layer neurons are generally chosen starting with lower number neurons and varying by trial and error till the configuration that gives minimum root-mean-square error is attained. The output (third or fourth layer) layer consists of one neuron of either onset date or length of critical dry spell. Two hidden layer networks may provide more benefits for some type of problems [60]. Several authors addressed this problem and considered more than one hidden layer (usually two hidden layers) in their network design processes.

A cross-correlation analysis was performed to measure the relationship between the predictors (attributes) and the predictands (classes). Positive and negative relationships were observed, some with weak relationships. Based on the cross-correlations, seven different ANN models shown in **Table 2** below are put forward for each station with a view to measuring their predictive ability by

Model	No. of input variable	Max. Temp	Min. Temp	Mean Temp	R. H. (06GMT)	R. H. (15GMT)	Wind Speed (2 m)	Sun hr.	Net Rad. (Rn)	Ref. evap. (ETo)
M1	9	✓	✓	✓	✓	✓	✓	✓	✓	✓
M2	8	✓	✓	✓	✓	—	✓	✓	✓	✓
M3	7	✓	✓	✓	✓	✓	✓	—	✓	—
M4	6	✓	✓	—	✓	—	✓	—	✓	✓
M5	5	✓	✓	—	✓	—	✓	—	✓	—
M6	4	✓	✓	—	—	—	✓	—	✓	—
M7	3	✓	—	—	—	—	✓	—	✓	—

Table 2.
Input variables used in the attempt to get suitable models (M1–M7) for the prediction of onset dates and lengths of critical dry spells.

comparing predicted values with the observed ones. A measure of ANN most suitable model performance on the basis of all statistical measures of the observed and predicted critical dry spell onset dates and lengths for the nine stations are shown in **Tables 3** and **4**. Out of the seven models used with these predictors, Model 1 having nine parameters was noted to be quite suitable for most of the stations, while models 2, 3, and 5 having eight, seven, and five parameters respectively were more suitable than Model 1 in some cases. Predictions (testings) were made for two regular mid-growing season critical dry spells (first and second) for all stations. Predictions were made on yearly basis on the twentieth (20th) day after the onset

Sta. name	Most suitable model (M) for dry spell onset	Neural net. Arch.	Lead time pred. Range (days)	RMSE	R^2	NSE	WIA	RSR	Prediction error margin (days)
Cal	M1-On Date1	9–3-1	27–34	1.53	0.75	0.72	0.89	0.50	−3.09 to 3.22
	M1-On Date2	9–9-1	41–56	2.93	0.82	0.82	0.95	0.40	−4.56 to 4.89
War	M1-On Date1	9–8–2-1	31–43	2.95	0.86	0.77	0.92	0.47	−4.67 to 3.01
	M1-On Date2	9–9–2-1	43–66	3.31	0.83	0.80	0.95	0.42	−1.08 to 4.75
Iba	M3-On Date1	7–9–2-1	15–26	1.42	0.80	0.64	0.93	0.57	−1.46 to 1.91
	M2-On Date2	8–9–9-1	28–37	2.07	0.70	0.65	0.90	0.56	−3.45 to 2.64
Ilo	M1-On Date1	9–8–1-1	17–24	1.65	0.70	0.68	0.91	0.53	−3.18 to 2.23
	M1-On Date2	9–8–1-1	30–37	1.54	0.86	0.85	0.96	0.37	−2.69 to 3.23
Lok	M1-On Date1	9–8–6-1	13–32	2.19	0.79	0.79	0.94	0.44	−3.07 to 3.89
	M5-On Date2	5–2-1	28–36	2.05	0.58	0.48	0.88	0.69	−3.55 to 3.15
Mak	M1-On Date1	9–9–8-1	20–24	1.12	0.75	0.59	0.82	0.61	−1.47 to 2.04
	M1-On Date2	9–8–4-1	26–44	2.70	0.79	0.76	0.94	0.46	−3.64 to 4.02
Yel	M1-On Date1	9–8–7-1	18–22	2.40	0.71	0.66	0.91	0.55	−4.04 to 3.42
	M1-On Date2	9–3-1	28–41	3.07	0.72	0.72	0.91	0.50	−4.29 to 4.07
Kad	M1-On Date1	9–9–2-1	13–28	2.67	0.71	0.65	0.91	0.56	−3.85 to 3.11
	M2-On Date2	8–7–1-1	26–34	2.37	0.79	0.74	0.93	0.48	−4.24 to 3.24
Yol	M1-On Date1	9–7–1-1	13–25	3.16	0.67	0.57	0.90	0.62	−3.40 to 2.63
	M1-On Date2	9–8–8-1	23–35	2.46	0.73	0.64	0.90	0.57	−4.08 to 4.24

Table 3.
A measure of ANN most suitable model performance on the basis of all statistical measures of the observed and predicted critical dry spell onset dates for the nine stations.

Agro-eco. zones	Sta. name	Most suitable model (M) for dry spell length	Neural net. arch.	RMSE	R^2	NSE	WIA	RSR	Prediction error margin (days)
Mangrove Swamp	Cal	M1-Len1	9–2–1	0.74	0.81	0.77	0.94	0.45	−0.97 to 1.27
		M5-Len2	5–3–1	0.97	0.69	0.67	0.87	0.55	−1.43 to 1.30
	War	M3-Len1	7–3–1	0.93	0.85	0.82	0.94	0.40	−1.32 to 1.51
		M1-Len2	9–5–4–1	0.46	0.92	0.69	0.94	0.53	−0.01 to 1.00
Rain Forest	Iba	M1-Len1	9–9–1–1	1.49	0.78	0.57	0.86	0.65	−1.71 to 2.02
		M1-Len2	9–4–1	1.52	0.75	0.57	0.79	0.62	−2.76 to 2.01
Southern Guinea Savannah	Ilo	M1-Len1	9–4–1	2.05	0.68	0.51	0.84	0.66	−3.38 to 2.68
		M2-Len2	8–4–1	0.79	0.93	0.90	0.98	0.30	−2.05 to 1.19
	Lok	M1-Len1	9–2–1	1.72	0.78	0.58	0.83	0.61	−1.45 to 2.91
		M1-Len2	9–2–1	1.61	0.82	0.55	0.87	0.64	−0.84 to 2.90
	Mak	M1-Len1	9–5–2–1	1.83	0.63	0.55	0.86	0.63	−2.27 to 2.41
		M1-Len2	9–5–4–1	1.83	0.71	0.63	0.91	0.58	−3.51 to 3.22
Northern Guinea Savannah	Yel	M1-Len1	9–8–1–1	1.71	0.78	0.64	0.89	0.59	−2.71 to 3.24
		M1-Len2	9–3–1	0.90	0.67	0.66	0.89	0.55	−1.33 to 1.51
	Kad	M1-Len1	9–3–1	1.85	0.76	0.58	0.83	0.61	−3.72 to 1.03
		M1-Len2	9–3–1	0.68	0.89	0.83	0.94	0.39	−1.34 to 0.91
	Yol	M1-Len1	9–3–1	1.65	0.71	0.67	0.87	0.55	−3.47 to 2.48
		M5-Len2	5–2–1–1	0.93	0.75	0.62	0.87	0.59	−0.93 to 1.93

Cal, War, Iba, Ilo, Lok, Mak, Yel, Kad, Yol—*Calabar, Warri, Ibadan, Ilorin, Lokoja, Makurdi, Yelwa, Kaduna, Yola;* ***M1 .. M7***—*Model1 .. Model7;* ***On Date1, On Date2***—*First Onset Date, Second Onset Date;* ***Len1, Len2***—*First Spell Length, Second Spell Length;* ***RMSE***—*Root-Mean-Square Error;* ***R2***—*Coefficient of Determination;* **NSE**—*Nash-Sutcliffe Coefficient of Efficiency;* ***WIA***—*Wilmott's Index of Agreement;* ***RSR***—*RMSE-Observations Standard Deviation Ratio.*

Table 4.
A measure of ANN most suitable model performance on the basis of all statistical measures of the observed and predicted critical dry spell lengths for the nine stations.

dates of growing season for Calabar and Warri with use of 10-day average values of the attributes (predictors), that is, average taken from the eleventh (11th) day through twentieth (20th) day. However, for the remaining seven stations, yearly predictions were made for the first and second critical dry spell onset dates and lengths on the thirtieth (30th) day after the onset dates of growing season with the use of 10-day average values of the attributes (predictors), that is, average taken from the twenty-first (21st) day through the thirtieth (30th) day. The choice of the 10-day average of the predictors and the choice of the beginning date were basically by trial and error until good predictors were realized. The predictions made for the onset dates of critical dry spells were actually for the number of days before the occurrence of first and second critical dry spells from the 20th or 30th day after the onset dates of growing season. So, the onset dates of the critical dry spells in terms of days of the year should be onset dates of growing season (in days of the year) plus 20 days (for Calabar and Warri) or 30 days (for the remaining seven stations) plus the number of days before the occurrence of the critical dry spell. These are indicated in Eqs. (2) and (3) respectively below:

For Calabar and Warri Stations (two station):

$$CDS_{OD} \text{ (day of year)} = OGS \text{ (day of year)} + 20 + NoD \tag{2}$$

For Ibadan, Ilorin, Lokoja, Makurdi, Yelwa, Kaduna, and Yola Stations (seven stations):

$$CDS_{OD}\ (\text{day of year}) = OGS\ (\text{day of year}) + 30 + NoD \tag{3}$$

where CDS_{OD} (day of year)—Critical Dry Spell Onset Date in days of year, OGS (day of year)—Onset Date of Growing Season in days of year, and NoD—number of days before the occurrence of the critical dry spell.

For any year, the normalized attributes (predictors) were substituted in the prediction equations (not shown) derived from the neural network architecture involving the input (predictors), hidden, and output (predictands) neurons. Normalized values of the predictors were used in the equation to limit the output to a range between 0 and 1, making the function useful in the prediction of probabilities. Outputs of hidden layer neurons and output layer neuron were determined using Sigmoid Activation Function. The purpose of the sigmoid activation function is to introduce nonlinearity into the output of a neuron. Neural network has neurons that work in correspondence of weight, bias, and activation function. After prediction, the predicted values were converted to actual values by the removal of the normalization.

3. Results and discussions

3.1 Warri and Calabar

Figure 3(a) and **(b)** gives the *first and second* respectively yearly actual and predicted values of mid-season critical dry spell onset dates and lengths for Warri in the Mangrove Swamp agro-ecological zone. The actual and predicted values of the onset dates of first and second critical dry spells are actually the *number of days* before the occurrence of the critical dry spells *from 20th day* after the onset dates of growing season. So, the predicted onset dates of the critical dry spells in terms of days of the year should be onset dates of growing season (in days of the year) plus 20 days plus the predicted number of days before the occurrence of the critical dry spell (Eq. (2)). The prediction lead times for first and second critical dry spell onset dates range from 31 to 66 days for Warri and from 27 to 56 days for Calabar (figure not shown) as shown in **Table 3**. The range of errors during testing for onset dates and lengths for the first and second critical dry spells is generally ±4 days for Warri and for Calabar (figure not shown). The root-mean-square errors (RMSE), coefficient of determination (R^2), Nash-Sutcliffe Coefficient of Efficiency (NSE), Wilmott's Index of Agreement (WIA), and RMSE-Observations Standard Deviation Ratio (RSR) for first and second critical dry spell *onset dates* for Warri range from 2.95 to 3.31, 0.83 to 0.86, 0.77 to 0.80, 0.92 to 0.95, and 0.42 to 0.47 days, respectively, while those for Calabar (figure not shown) range from 1.53 to 2.93, 0.75 to 0.82, 0.72 to 0.82, 0.89 to 0.95, and 0.40 to 0.50 days, respectively (**Table 3**). The RMSE, R^2, NSE, WIA, and RSR for the first and second critical dry spell *lengths* for Warri range from 0.46 to 0.93, 0.85 to 0.92, 0.69 to 0.82, 0.94 to 0.94, and 0.40 to 0.53 days, respectively, while those for Calabar (figure not shown) range from 0.74 to 0.97, 0.69 to 0.81, 0.67 to 0.77, 0.87 to 0.94, and 0.45 to 0.55 days, respectively (**Table 4**). The neural network architecture for the first and second onset dates and lengths of the critical dry spells for both stations are given in **Tables 3** and **4**.

Ogunrinde et al. [3] who applied ANN for forecasting Standardized Precipitation and Evapotranspiration Index (SPEI): A case study of Nigeria got an RMSE value of 0.7476 for Ikeja, a station in Lagos State, Western Nigeria in the same agro-ecological

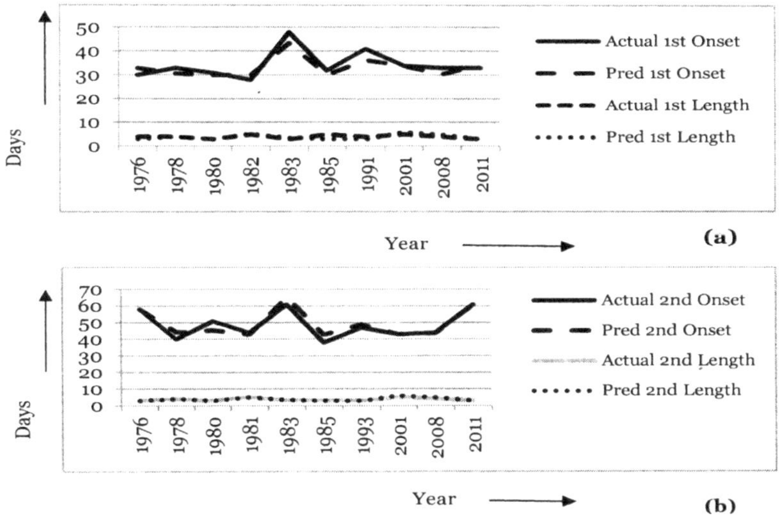

Figure 3.
Yearly prediction of (a) mid-season first critical dry spell onset dates and lengths (b) mid-season second critical dry spell onset dates and lengths in Warri. The critical dry spell onset dates are given in terms of the number of days before the critical dry spell occurrence from the 20th day after the onset dates of growing season.

zone with Warri and Calabar. The two values are somehow close especially for the initial aspect of the range for the critical dry spell lengths even though the forecasts in the current work are for the onset dates and lengths of mid-season critical dry spells and not for SPEI. Dry spell onset dates and lengths prediction were not carried out by these and other researchers. So, on the 20th day after the onset date of growing season of any year in Warri and Calabar, maize farmers could be given yearly advance information on the dates of occurrence of first and second critical dry spells and their respective lengths for the mid-season (tasseling and flowering of maize). This would enable them make adequate preparations for their farming operations to ensure improved maize yield taking cognizance of the error margins.

3.2 Ibadan

The yearly actual and predicted values of the *first and second* mid-season critical dry spell onset dates and lengths are shown in **Figure 4(a)** and **(b)** respectively for Ibadan in the Rain Forest agro-ecological zone. The actual and predicted values of the onset dates of first and second critical dry spells are actually the *number of days* before the occurrence of the critical dry spells *from 30th day* after the onset dates of growing season. So, the actual and predicted onset dates of the critical dry spells in terms of days of the year should be onset dates of growing season (in days of the year) plus 30 days plus the predicted number of days before the occurrence of the critical dry spell (Eq. (3)). The first and second critical dry spell onset date prediction lead times range from 15 to 37 days in Ibadan (**Table 3**). The range of errors during testing for onset dates and lengths for the first and second critical dry spells is generally ±4 days. The RMSE, R^2, NSE, WIA, and RSR for first and second critical dry spell *onset dates* for Ibadan range from 1.42 to 2.07, 0.70 to 0.80, 0.64 to 0.65, 0.90 to 0.93, and 0.56 to 0.57 days, respectively (**Table 3**), while those for first and second critical dry spell *lengths* range from 1.49 to 1.52, 0.75 to 0.78, 0.57 to 0.57, 0.79 to 0.86, and 0.62 to 0.65 days, respectively (**Table 4**). The neural network architecture for first and second onset dates and lengths of the critical dry spells for Ibadan are also given in **Tables 3** and **4**.

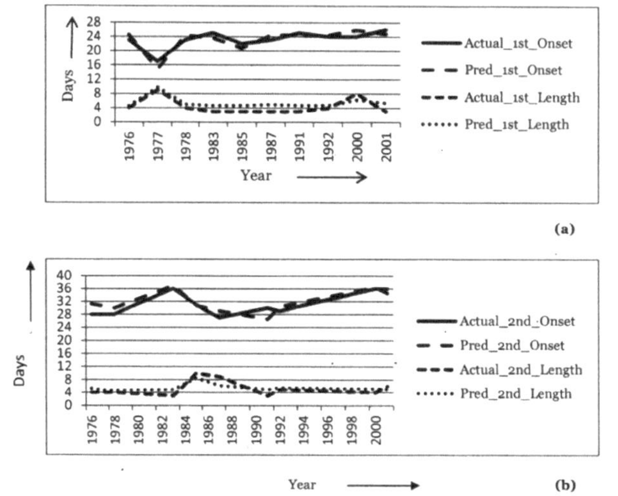

Figure 4.
Yearly prediction of (a) mid-season first critical dry spell onset dates and lengths (b) mid-season second critical dry spell onset dates and lengths in Ibadan. The critical dry spell onset dates are given in terms of the number of days before the critical dry spell occurrence from the 30th day after the onset dates of growing season.

The result of the work of Morid et al. [15] regarding ANN forecast of Effective Drought Index (EDI) (6 months in advance) in Mehrabad station using nine input models gave validation RMSE values ranging from 0.55 to 1.51. Though the latitudes of Ibadan and Mehrabad differ and the target forecasts also differ, the range of RMSE values is somewhat close. The predictions of critical dry spell onset dates and lengths were not addressed by the researchers.

Therefore, on the 30th day after the onset date of growing season of any year in the station, maize farmers could be given advance information on the dates of occurrence of first and second critical dry spells and their respective lengths to enable them make informed preparations in their farming operations for enhanced maize yield taking note of the error margins.

3.3 Makurdi, Ilorin, and Lokoja

Figure 5(a) and **(b)** shows the *first* and *second* respectively yearly actual and predicted values of mid-season critical dry spell onset dates and lengths for Makurdi in the Southern Guinea Savannah agro-ecological zone. The actual and predicted values of the onset dates of first and second critical dry spells are actually the *number of days* before the occurrence of the critical dry spells *from 30th day* after the onset dates of growing season. So, the predicted onset dates of the critical dry spells in terms of days of the year should be onset dates of growing season (in days of the year) plus 30 days plus the predicted number of days before the occurrence of the critical dry spell (Eq. (3)). The first and second critical dry spell onset date prediction lead times range from 20 to 44 days for Makurdi, those for Ilorin (figure not shown) range from 17 to 37 days, while those for Lokoja (figure not shown) range from 13 to 36 (**Table 3**). The range of errors during testing for onset dates and lengths of the first and second critical dry spells is generally ±4 days for each of the stations—Makurdi, Ilorin (figure not shown), and Lokoja (figure not shown). The RMSE, R^2, NSE, WIA, and RSR for first and second critical dry spell *onset dates* for Makurdi range from 1.12 to 2.70, 0.75 to 0.79, 0.59 to 0.76, 0.82 to 0.94, and 0.46 to 0.61 days, respectively, those for Ilorin (figure not shown) range from 1.54 to 1.65, 0.70 to 0.86, 0.68 to 0.85, 0.91 to 0.96 and 0.37 to 0.53 days,

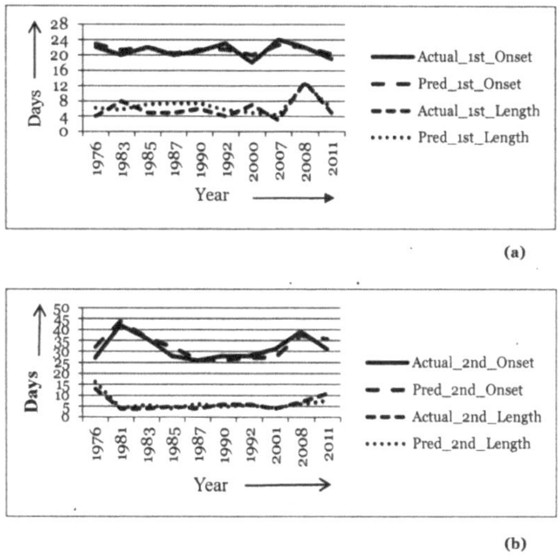

Figure 5.
Yearly prediction of (a) mid-season first critical dry spell onset dates and lengths (b) mid-season second critical dry spell onset dates and lengths respectively in Makurdi. The critical dry spell onset dates are given in terms of the number of days before the critical dry spell occurrence from the 30th day after the onset dates of growing season.

respectively, while those for Lokoja (figure not shown) range from 2.05 to 2.19, 0.58 to 0.79, 0.48 to 0.79, 0.88 to 0.94, and 0.44 to 0.69 days, respectively(**Table 3**). The RMSE, R^2, NSE, WIA, and RSR for first and second critical dry spell *lengths* for Makurdi range from 1.83 to 1.83, 0.63 to 0.71, 0.55 to 0.63, 0.86 to 0.91, and 0.58 to 0.63 days, respectively; those for Ilorin (figure not shown) range from 0.79 to 2.05, 0.68 to 0.93, 0.51 to 0.90, 0.84 to 0.98, and 0.30 to 0.66 days, respectively, while those for Lokoja (figure not shown) range from 1.61 to 1.72, 0.78 to 0.82, 0.55 to 0.58, 0.83 to 0.87, and 0.61 to 0.64 days, respectively (**Table 4**). The neural network architecture for first and second onset dates and lengths of the critical dry spells for the three stations are given in **Tables 3** and **4**.

Ogunrinde et al. [3] in their work on ANN for forecasting SPEI: A case study of Nigeria (for drought matters) got an RMSE value of 0.5957 for Lokoja station. The difference in the values got for critical dry spell length in the present work is possibly as a result of different target forecasts—SPEI as distinct from mid-season critical dry spell lengths. Therefore, on the 30th day after the onset date of growing season of any year in the stations, maize farmers could be given yearly advance information on the dates of occurrence of first and second critical dry spells and their respective lengths. This would enable farmers make necessary plans for their farming operations for enhanced maize yield noting the prediction error margins.

3.4 Kaduna, Yelwa, and Yola

The yearly actual and predicted values of *first* and *second* mid-season critical dry spell onset dates and lengths are presented in **Figure 6(a)** and **(b)** respectively for Kaduna in Northern Guinea Savannah agro-ecological zone. The actual and predicted values of the onset dates of first and second critical dry spells are actually the *number of days* before the occurrence of the critical dry spells *from 30th day* after the onset dates of growing season. So, the predicted onset dates of the critical dry spells in terms of days of the year should be onset dates of growing season (in days of the year) plus 30 days plus the predicted number of days before the occurrence of the critical dry spell (Eq. (3)). The prediction lead times for first and second

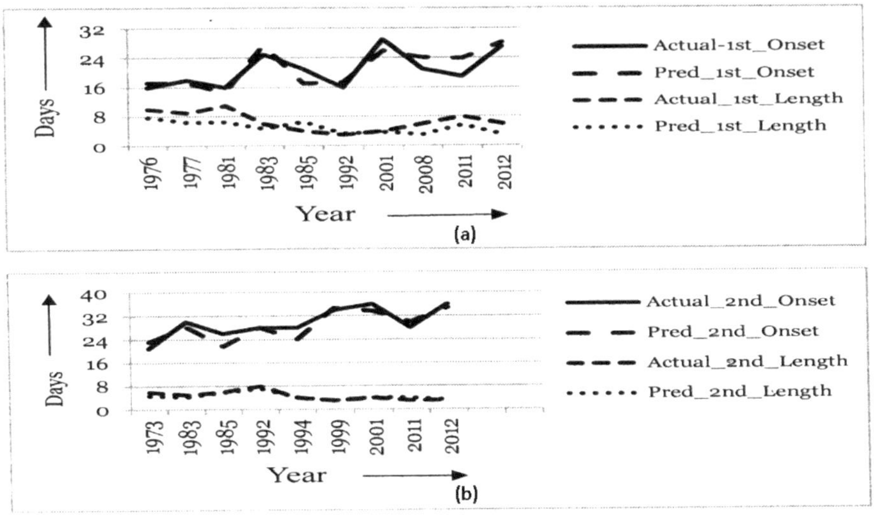

Figure 6.
Yearly prediction of (a) mid-season first critical dry spell onset dates and lengths (b) mid-season second critical dry spell onset dates and lengths respectively in Kaduna. The critical dry spell onset dates are given in terms of the number of days before the critical dry spell occurrence from the 30th day after the onset dates of growing season.

critical dry spells range from 13 to 34 days for Kaduna. Those for Yelwa (figure not shown) range from 18 to 41 days, while those for Yola (figure not shown) range from 13 to 35 days (**Table 3**). The range of errors during testing for onset dates and lengths of the first and second critical dry spells is generally ±4 days for each of the stations—Kaduna, Yelwa (figure not shown), and Yola (figure also not shown). The RMSE, R^2, NSE, WIA, and RSR for *onset dates* of first and second critical dry spells for Kaduna range from 2.37 to 2.67, 0.71 to 0.79, 0.65 to 0.74, 0.91 to 0.93, and 0.48 to 0.56 days, respectively. Those for Yelwa (figure not shown) range from 2.40 to 3.07, 0.71 to 0.72, 0.66 to 0.72, 0.91 to 0.91, and 0.50 to 0.55 days, respectively, while those for Yola (figure not shown) range from 2.46 to 3.16, 0.67 to 0.73, 0.57 to 0.64, 0.90 to 0.90, and 0.57 to 0.62 days, respectively (**Table 3**). The RMSE, R^2, NSE, WIA, and RSR for first and second critical dry spell *lengths* for Kaduna range from 0.68 to 1.85, 0.76 to 0.89, 0.58 to 0.83, 0.83 to 0.94, and 0.39 to 0.61 days, respectively, those for Yelwa (figure not shown) range from 0.90 to 1.71, 0.67 to 0.78, 0.64 to 0.66, 0.89 to 0.89, and 0.55 to 0.59 days, respectively, while those for Yola (figure not shown) range from 0.93 to 1.65, 0.71 to 0.75, 0.62 to 0.67, 0.87 to 0.87, and 0.55 to 0.59 days, respectively (**Table 4**). The neural network architecture for the first and second onset dates and lengths of the critical dry spells for the three stations are also given in **Tables 3** and **4**.

Mulualem and Liou [14] in their work on ANN in forecasting SPEI for the Upper Blue Nile Basin in Ethiopia got RMSE value of 0.428 for Bahdir Dar of almost the same latitude with Yelwa, Nigeria. The value of 0.91–1.71 got for critical dry spell length got in the current work is somehow close. However, the difference in the values could be as a result of different target forecasts—SPEI as distinct from mid-season critical dry spell lengths. Therefore, on the 30th day after the onset date of growing season of any year in the stations, maize farmers could be given advance information on the dates of occurrence of first and second critical dry spells and their respective lengths to enable them make adequate plans for their farming operations for improved maize yield. The prediction could be made using ANN on 30th day after growing season onset dates for these critical dry spells with 10-day average values of the predictors (attributes) taken from 21st to 30th day with minimum lead times of about 2 weeks and maximum of about 2 month as given above (**Table 3**). To make predictions for any year, the predictors (attributes) are first normalized and substituted into the equation (not shown) derived from the neural network architecture involving the input, hidden and output layers, weights, and sigmoid activation functions. At the result stage, the normalization is removed to get the actual onset dates and lengths of the critical dry spells—predictands (classes).

4. Conclusions and recommendation

The prediction of mid-season critical dry spell onset dates and lengths for 118 day rain-fed maize crop in Nigeria using ANN has yielded the following useful results that include the following: (a) the provision of yearly advance information on the number of days before the occurrence of first and second critical dry spells and their respective lengths on 20th day after the onset dates of growing season in Calabar and Warri; (b) the provision of yearly advance information on the number of days before the occurrence of first and second critical dry spell onset dates and lengths on 30th day after the onset dates of growing season in Ibadan, Ilorin, Lokoja, Makurdi, Yelwa, Kaduna, and Yola in Nigeria. The minimum prediction lead time is about 2 weeks, while the maximum is about 2 months. This information will aid yearly supplementary irrigation planning, scheduling, and various other decision makings related to sustainable agricultural operations for enhanced 118-day maize yield in the nine stations in Nigeria. For future work, it is recommended that more stations and longer years of data be used to ensure adequate training of ANN networks to realize better prediction results and gain more insight into dry spell occurrences during mid-growing seasons in Nigeria.

Acknowledgements

We are grateful to Dr. Imoleayo Gbode for his technical assistance regarding the downloads and use of neuralnet package for this work. We are indebted to the former Director General/Chief Executive Officer of Nigerian Meteorological Agency (NiMet) and Dr. A. C. Anuforom, for the provision of the data used for this work. We thank immensely the former Director of West African Science Center for Climate Change and Adapted Land Use (WASCAL) of The Federal University of Technology, Akure, Nigeria, Prof. K. O. Ogunjobi, for granting permission for the use of its Internet facility. We place on record the technical assistance of Dr. O. Adeyeri on matters of data homogenization. The Director of the Centre for Continuing Education, The Federal University of Technology, Akure, Prof. E. C. Okogbue is highly appreciated for his good advice in this work and for the provision of some academic facility used for this work.

Conflict of interest

The authors declare that they have no conflict of interest as regards the publication of this article.

Author details

Nnadozie Okonkwo Nnoli[1*], Ahmed Balogun[1], Jerome Omotosho[1] and Samuel Agele[2]

1 Department of Meteorology and Climate Science, The Federal University of Technology, Akure, Nigeria

2 Department of Crop, Soil and Pest Management, The Federal University of Technology, Akure, Nigeria

*Address all correspondence to: nnolino@futa.edu.ng; nonnoli48@gmail.com

References

[1] Song Y, Tian J, Linderholm HW, Wang C, Ou Z, Chen D. The contributions of climate change and production area expansion to drought risk for maize in China over the last four decades. International Journal of Climatology. 2020;41(sup 1): E2851-E2862. DOI: 10.1002/joc.6885

[2] Yu X, He X, Zheng H, et al. (2013) Spatial and temporal analysis of drought risk during the crop-growing season over Northeast China. NatHazards. 2013;**71**:275-289

[3] Ogunrinde AT, Oguntunde PG, Fasinmirin JT, Akinwumiju AS. Application of artificial neural network for forecasting standardized precipitation and evapo-transpiration index: A case study of Nigeria. Engineering Reports. 2020;**2020**:e12194. DOI: 10.1002/eng2.12194 wileyonlinelibrary.com/journal/eng2

[4] Mugalavai EM, Kipkorir EC, Songok CK. Evaluation of dry spells during sensitive growth stages for maize crop in Western Kenya. In: The International Conference on Disaster Risk Reduction and Conflict Resolution for Sustainable Development 18-20th July, 2012. Kakamega, Kenya: @ Mmust; 2012

[5] Gao C, Li X, Sun Y, et al. 2019 Water requirement of summer maize at different growth stages and the spatiotemporal characteristics of agricultural drought in the Huaihe River Basin, China. Theoretical and Applied Climatology. 2019;**136**:1289-1302. DOI: 10.1007/s00704-018-2558-6

[6] Allen RG, Pereira LS, Raes D, Smith M. FAO Crop Evapotranspiration - Guidelines for Computing Crop Water Requirements; FAO Irrigation and Drainage Paper 56. Rome, Italy; 1998:1-300

[7] Sharma TC. Challenges in drought research: Some perspectives and future directions. Hydrological Science Journal. 2002;**47**(sup 1):S19-S30. DOI: 10.1080/02626660209493019

[8] IPCC. Special Report on Managing the Risk of Extreme Events and Disasters to Advance Climate Change Adaptation. Cambridge, UK: Cambridge University Press; 2012. p. 582

[9] Wilhite DA, Sivakumar MVK, Pulwarty R. Managing drought in a changing climate: The role of national drought policy weather and climate. Extremes. 2014;**3**(2014):4-13

[10] Box GEP, Jenkins GM. Time Series Analysis: Forecasting and Control. San Francisco, Calif, USA: Holden-Day; 1976

[11] Gbangou T, Ludwig F, van Slobbe E, Greuell W, Kranjac-Berisavljevic G. Rainfall and dry spell occurrence in Ghana: Trends and seasonal predictions with a dynamical and a statistical model. Theoretical and Applied Climatology. 2020;141:371-387. DOI: 10.1007/s00704-020-03212-5

[12] Mishra SS, Nagarajan R. Forecasting drought in Tel River Basin using feed-forward recursive neural network. In: 2012 International Conference on Environmental, Biomedical and Biotechnology IPCBEE vol. 41 (2012) © (2012). Singapore: IACSIT Press; 2012

[13] Abrishami N, Sepaskhah AR, Shahrokhnia MH. Estimating wheat and maize daily evapotranspiration using artificial neural network. Theoretical and Applied Climatology 2019;135: 945-958. https://doi.org/10.1007/s00704-018-2418-4

[14] Mulualem GM, Liou Y-A. Application of Artificial Neural Networks in Forecasting a Standardized Precipitation Evapotranspiration Index for the Upper Blue Nile Basin. Water. 2020;**12**(643):1-19. DOI: 10.3390/

w12030643www.mdpi.com/journal/water

[15] Morid S, Smakhtin V, Bagherzadeh K. (2007) Drought forecasting using artificial neural networks and time series of drought indices. International Journal of Climatology. 2007;**27**:2103-2111

[16] Weerasinghe HDP, Premaratne HL, Sonnadara DUJ. Performance of neural networks in forecasting daily precipitation using multiple choices. J. Natn. Sci. Foundation Sri Lanka. 2010;**38**(3):163-170

[17] Mathugama SC, Peiris TSG. Critical evaluation of dry spell research. International Journal of Basic and Applied Sciences IJBAS-IJENS. 2011;**11**(06):153-160

[18] NiMet NWP Unit with GCRF African SWIFT Project. Sub Seasonal - to- Seasonal (S2S) CLIMATE Forecast over West African Sub-Region with Special Emphasis on Nigeria. Abuja, Nigeria: NiMet-SWIFT Publication; 2020. pp. 1-11

[19] Nigerian Meteorological Agency. Overview of the 2020 Seasonal Rainfall Prediction, (prepared by J. I. Adamu). Abuja: Nigerian Meteorological Agency Publication; 2020. pp. 1-40

[20] Dee DP, Uppala SM, Simmons AJ, Berrisford P, Poli P, Kobayashi S, et al. The ERA-Interim reanalysis: Configuration and performance of the data assimilation system. Quart. J. Roy. Meteorol. Soc. 2011;**137**(656):553-597

[21] Huffman GJ, Bolvin DT, Nelkin EJ, Wolff DB, Adler RF, Gu G, et al. The TRMM multisatellite precipitation analysis (TMPA): quasi-global, multiyear, combined-sensor precipitation estimates at fine scales. Journal of Hydrometeorology. 2007;**8**(1):38-55

[22] Domonkos P, Coll J. Homogenisation of temperature and precipitation time series with ACMANT3: method description and efficiency tests. International Journal of Climatology. 2017;**37**:1910-1921

[23] Adeyeri OE, Laux P, Lawin AE, Ige SO, Kunstmann H. Analysis of hydrometeorological variables over the transboundary Komadugu-Yobe basin. West Africa: Journal of Water and Climate Change. 2020;1339-1354. Doi: 10.2166/wcc.2019.283. (http://creativecommons.org/licenses/by/4.0/)

[24] Benoit P. The start of the growing season in Northern Nigeria. Agricultural Meteorology. 1977;**18**:91-99

[25] Odekunle TO. Determining rainy season onset and retreat over Nigeria from precipitation amount and number of rain days. Theoretical and Applied Climatology. 2006;**83**:163-201

[26] Nema MK, Khare D, Chandniha SK. Application of artificial intelligence to estimate the reference evapotranspiration in sub-humid Doon valley. Applied Water Science. 2017;7:3903-3910. DOI: 10.1007/s13201-017-0543-3

[27] Djaman K, O'Neill M, Diop L, et al. 2019 Evaluation of the Penman-Monteith and other 34 reference evapotranspiration equations under limited data in a semiarid dry climate. Theoretical and Applied Climatology. 2019;**137**:729-743. DOI: doi. 10.1007/s00704-018-2624-0

[28] Paredes P, Fontes JC, Azevedo EB. *et al.* 2018 Daily reference crop evapotranspiration in the humid environments of Azores islands using reduced data sets: accuracy of FAO-PM temperature and Hargreaves-Samani methods. Theoretical and Applied Climatology (2018);134:595-611. https://doi.org/10.1007/s00704-017-2295-2.

[29] Douguedroit A. 1987 The variations of dry spells in marseilles from 1865 to 1984. International Journal of Climatology. 1987;7:541-551

[30] Raymond F, Ullmann A, Camberlin P, Drobinski P, Chateau Smith C. Extreme dry spell detection and climatology over the Mediterranean Basin during the wet season. Geophysical Research Letters. 2016;**43**:7196-7204. DOI: 10.1002/2016GL069758

[31] Raymond F, Ullmann A, Camberlin P, Oueslati B, Drobinsky P. 2018 Atmospheric conditions and weather regimes associated with extreme winter dry spells over the Mediterranean basin. Clim. Dynam. 2018;**50**:4437-4453. DOI: 10.1007/s00382- 017-3884-6

[32] Serra C, Lana X, Burgueno A, Martinez MD. 2016 Partial duration series distributions of the European dry spell lengths for the second half of the twentieth century. Theoretical and Applied Climatology. 2016;**123**:63-81

[33] Tramblay Y, Hertig E. 2018 Modelling extreme dry spells in the Mediterranean region in connection with atmospheric circulation. Atmospheric Research. 2018;**202**:40-48

[34] Lobell DB, Hammer GL, Chenu K, Zheng B, Mclean G, Chapman SC. 2015 The shifting influence of drought and heat stress for crops in northeast Australia. Global Change Biology. 2015;**21**:4115-4127. DOI: doi.org/10.1111/gcb.13022

[35] Allen CD, Breshears DD, McDowell NG. On underestimation of global vulnerability to tree mortality and forest die off from hotter drought in the Anthropocene. Ecosphere. 2015;**6**(8):1-55. DOI: 10.1890/ES15-00203.1

[36] Anderegg WRL, Klein T, Bartlett M, Sack L, Pellegrini AFA, Choat B, et al. 2016 Meta-analysis reveals that hydraulic traits explain cross-species patterns of drought-induced tree mortality across the globe. P. Natl. Acad. Sci. USA. 2016;**113**:5024-5029

[37] McMahon TA, Peel MC, Lowe L, Srikanthan R, McVicar TR. 2013 Estimating actual, potential, reference crop and pan evaporation using standard meteorological data: a pragmatic synthesis. Hydrology and Earth System Sciences. 2013;**17**:1331-1363. DOI: doi.org/10.5194/hess-17-1331-2013

[38] Rivoire P, Tramblay Y, Neppel L, Hertig E, Vicente-Serrano SM. Impact of the dry-day definition in Mediterranean extreme dry-spell analysis Nat. Hazard Earth Syst. Sci. 2019;**19**:1629-1638. DOI: hess-19-1629-2019

[39] Engelbrecht BMJ, Dalling JW, Perason TRH, Wolf RL, et al. Short dry spells in the wet season increase mortality of tropical pioneer seedlings. Oecologia. 2006;**148**(2):258-269

[40] Sawa BA, Adebayo AA. Effects of Pentad dry spells on the yield of some crops in the semi-arid eco-climate region of Northern Nigeria. The Zaria Geographer. 2018;**19**(1):49-60 [Accessed: 12 February 2020]

[41] Luk KC, Ball JE, Sharma A. An application of neural networks for rainfall forecasting. Mathematical and Computer Modelling. 2001;**33**(6-7):683-693

[42] Zhang G, Patuwo BE, Hu MY. Forecasting with artificial neural network: The state of the art. International Journal of Forecasting. 1998;**14**(1998):35-62. DOI: 10.1016/S0169-2070(97)00044-7

[43] Ripley BD. Statistical aspects of neural networks network and chaos - statistical and probabilistic aspects

Chapter 2. In: Barn-dorff-Nielsen OE, Jensen JL and Kendall WS, editors. Chapman k Hall, London: Networks and Chaos: Statistical and Probabilistic Aspects. 1993. pp. 40-123

[44] Cheng B, Titterington DM. Neural Networks : A review from a statistical perspective. Statistical Science. 1994;**9**(1):2-30

[45] Hornik K, Stinchcombe M, White H. Multilayer feedforward networks are universal approximators. Neural Networks. 1989;**2**:359-366

[46] Hornik K. Some new results on neural network approximation. Neural Networks. 1993;**6**:1069-1072

[47] Cybenko G. Approximation by superpositions of a sigmoidal function. Math. Control Signal Systems. 1989;**2**:303-314. DOI: 10.1007/BF02551274

[48] Affandi AK, Watanabe K. Daily groundwater level fluctuation forecasting using soft computing technique. Nature and Science. 2007;**5**(2):1-10

[49] Gupta MM, Jin L and Homma, N. Static and Dynamic Neural Networks: From Fundamentals to Advanced Theory. Hoboken, New Jersey, John Wiley & Sons, Inc; 2003. pp. 1-722

[50] Haykin S. Neural Networks: A Comprehensive Foundation. 2nd. ed. Upper SaddleRever, New Jersey: Prentice Hall; 1999

[51] Akpan VA, Hassapis GD. Nonlinear model identification and adaptive model predictive control using neural networks. ISA Transactions. 2011;**50**(2):177-194. DOI: 10.1016/j.isatra.2010.12.007 Epub 2011 Feb 1. PMID: 21281932

[52] R Core Team. R: A Language and Environment for Statistical Computing. Vienna, Austria: R Foundation for Statistical Computing; 2017. Available from: https://www.R-project.org/

[53] Fritsch S and Guenther F. 2016. guenther@leibniz-bips.de, German Research Foundation (DFG: http://www.dfg.de) under grant scheme PI 345/3-1

[54] Riedmiller M. *Rprop - Description and Implementation Details*. Technical Report. Karlsruhe, Germany: University of Karlsruhe; 1994

[55] Riedmiller M, Braun H. A direct adaptive method for faster backpropagation learning: The RPROP algorithm. In: Proceedings of the IEEE International Conference on Neural Networks (ICNN). San Francisco, CA, USA: IEEE; 1993. pp. 586-591. DOI:10.1109/ICNN.1993.298623

[56] Anastasiadis AD, Magoulas GD, Vrahatis MN. *New globally convergent training scheme based on the resilient propagation algorithm*. Neurocomputing. 2005;**64**:253-270

[57] Vamsidhar E, Varma KVSRP, Rao PS, Satapati R. Predicting rainfall using backpropagation neural network model. International Journal on Computer Science and Engineering. 2010;**02**(04):1119-1121

[58] Dubey AD. Artificial neural network models for rainfall prediction in Pondicherry. International Journal of Computer Applications (0975-8887). 2015;**120**(3):30-35

[59] Kumar A, Kumar A, Ranjan R, Kumar S. A rainfall prediction model using artificial neural network. In: Control and Syst. Graduate Research Colloq. (ICSGRC). 2012. pp. 82-87

[60] Barron AR. A comment on "Neural networks: A review from a statistical perspective". Statistical Science. 1994;**9**(1):33-35

Chapter 7

Improving Maize Shelling Operation Using Motorized Mobile Shellers: *A Step towards Reducing Postharvest Losses in Low Developing Countries*

Denis Nsubuga, Isa Kabenge, Ahamada Zziwa, Nicholas Kiggundu, Joshua Wanyama and Noble Banadda

Abstract

Maize shelling is still a challenge in low developing countries with more efforts required to advance this operation. In Uganda, motorized immobile maize shellers have been fabricated locally to enhance the shelling operation. However, their performance has not elated the farmers. The unsatisfactory performance is a result of these shellers being fabricated by local artisan with finite understanding of the maize grain characteristics and operation factors to optimize maize shelling. In addition, farmers in these countries have a deficiency of power to operate the motorized maize shellers available. Transportation of these motorized maize shellers is also still a challenge and it imposes an extra cost to the farmers hence reducing their profits from maize growing. In this chapter, we reviewed maize shelling process in low developing countries particularly the categories of maize shelling, maize sheller design requirements, use of equations to design sheller parts, modification of the motorized maize shellers and case studies on the mobile maize shellers, comparing them with immobile maize shellers. The study concluded that on addition to other sheller performance attributes, motorized mobile maize shellers can solve transportation challenges associated with motorized immobile maize shellers.

Keywords: maize, shelling, mobile motorized shellers, post-harvest operations

1. Introduction

Maize is among the three critical cereal grains in the world, others being wheat and rice [1]. Maize was first identified in central Mexico 7000 years ago from a wild grass and Indigenous Americans converted it into food [2]. This cereal grain contains starch (60–80%), protein (8–12%), fat (3–5%), and minerals (1–2%) [3, 4]. It is grown worldwide, with Unites States, China, and Brazil as the top three maize-producing countries with a combined production of approximately 563 of the 717 million tons/year [2]. Maize contains nutrients for both humans and animals but it is also used for production of starch, oil and protein, alcoholic beverages, food sweeteners, and biofuels [5]. The significance of maize as a staple food in low developing countries can be compared to that of wheat in Asia. It is mostly consumed in Eastern, Western and Southern Africa in different forms such as *kenkey* in Ghana, *Ogi* in Nigeria, stiff porridge (*nsima*) in Malawi, maize meal (*ugali*) in Kenya [6], and posho and porridge in Uganda. In Sub Saharan Africa (SSA), over 208 million people bank on maize as a food source and being economically empowered [7]. Out of the 22 countries in the world where maize is mostly consumed, 16 of them are found in Africa [7]. This makes maize a very important cereal in Africa. Despite its importance, the losses of maize after harvest have decreased its availability among the poor people in Africa. In Uganda, for example, maize postharvest losses are about 30% [1] which has escalated hunger especially among the poor in the villages.

Maize processing include harvesting, dehusking, drying, shelling, storing, and milling. Compared to other operations, shelling still stand out as the most challenging operation that requires more work to improve it [8]. For the maize farmers to fully enjoy the financial benefits from their maize, appropriate technology that suits their needs is a requirement. In this regard, motorized immobile maize shellers have been fabricated locally to enhance the shelling operation. However, their performance has not elated the farmers. The unsatisfactory performance is a result of these shellers being fabricated by local artisans with finite understanding of the maize grain characteristics and operation factors to optimize maize shelling [1]. In addition, farmers in low developing countries have a deficiency of power to operate the motorized maize shellers available. It has been reported that transportation of these immobile maize shellers with the engines to run them from place to place is a big problem to sheller service providers; often requiring an additional carrier to move shellers to the farmers' field. The shelling service providers hence ask for an extra cost, which is usually passed on to farmers. These shellers also require extra time and energy to arrange the maize shelling environment at the farm level [9].

To consider the shelling power and sheller transportation problems, low cost motorized mobile maize shelling technologies have been developed as a result of modifying the available motorized immobile maize shellers. Some motorized mobile maize shellers were fabricated in 2012 by industrious fabricator Munyegera Agro-Machinery in Eastern part of Uganda [10]. Later, the multipurpose vehicle mobile maize shelling technology was introduced [1]. In Bangladesh, a two-wheel tractor mounted mobile sheller for small scale farmers was also introduced [9]. In this book chapter, maize shelling operation in low developing countries has been described with focus on encouraging a paradigm shift from the motorized immobile maize shellers to mobile maize shellers as a solution to the maize shelling constraints in these countries.

2. Maize shelling as a postharvest operation

Maize shelling as a postharvest operation is the removal of maize seeds from the cob [11]. This operation can be carried out either in the field or at the storage facility. Maize shelling is therefore an important step towards the processing of maize to various finished products like flour and maize bran.

2.1 Maize shelling in developed countries

In developed countries like Europe, North America, and China, maize shelling operation is done using combine harvesters [12]. Combine harvesters (**Figure 1**)

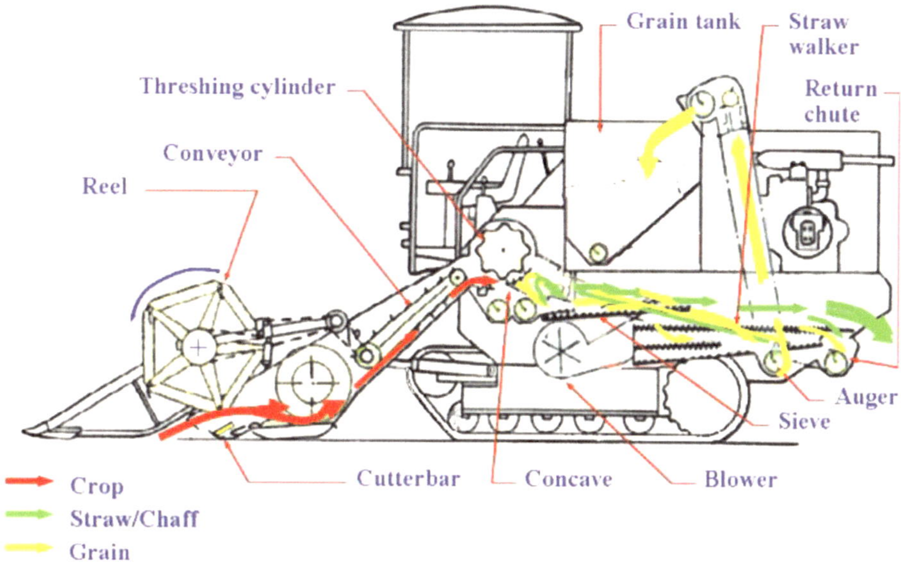

Figure 1.
Different components of a combine harvester [13].

simultaneously perform operations of ear picking, threshing, separation, and cleaning on the mature maize plants in the field. The purpose of this mechanized maize harvesting technology is to replace manual labor to harvest maize from fields in time with minimum loss while maintain high quality standards [14]. Some of the advantages of mechanized maize shelling include: reduced drudgery, enhanced productivity, time consciousness of agricultural operation, and availing labor for other agricultural operations. Combine harvester designers are working towards the quality of the process automatic controls and protecting the environment [15].

2.2 Maize shelling in low developing countries

Maize shelling in low developing countries is still a challenge to its value addition as it is tiresome and requires a number of labor hours [11]. A major issue for maize value chain is that good quality maize is difficult to find among farmers. Many times, buyers are ready to pay a high price for maize grains from farmers with good quality maize. However, good quality maize is often unavailable due to poor postharvest handling. The impacts of quality at postharvest level can be attributed to poor drying and storage methods among other factors. For example, maize drying on the bare ground, and storage in dump places and aflatoxin growth [10]. Beside drying and poor storage, maize post-harvest losses are also due to use of rudimentary tools like tapered cylindrical metallic shelling device [16].

Maize shelling methods can be categorized as traditional maize shilling, manual maize shelling, and motorized maize shelling based on the technology used.

2.2.1 Tradition maize shelling

Maize is shelled traditionally by hand (**Figure 2**). Here, the grains are detached from the cob by pressing them with the thumb [2]. The technique produces unbroken kernels but the process is tedious. A few kilograms can be shelled in an hour, with damages left on shellers' fingers. Another simple and common method of traditional maize shelling is to rub two maize cobs against each other in order to

Figure 2.
Maize shelling by hand [16].

detach the maize kernels [17]. However, these traditional methods of shelling are, not efficient, consume a lot of time, and require a lot of energy with very low productivity since farmers can shell only a few kilograms/hour.

2.2.2 Manual maize shelling

This method is almost similar to the traditional method of shelling except that it requires more energy compared to traditional methods to run manual maize sheller (**Figure 3**). For some manual shellers, two people are required during shelling, one person constantly feeds the maize cobs and the other operates the equipment by rotating the handle [8] while other manual shellers require one person [2]. Hand-operated shellers, requires less time to shell the maize compared to the traditional methods. These come in several models, and they are usually driven by rotating the handle or a pedal. With the output capacity of 14–100 kg/min, they are more suitable for small-scale maize production [2]. Hand-operated maize shellers are also suitable for shelling maize for seed purpose since damaged maize kernels are fewer compared to motorized maize shellers [18].

2.2.3 Motorized maize shellers

This method uses the same concept as hand-operated maize shellers except that the shellers are powered using a motor or an engine (**Figure 4**). The shellers under this method can be categorized into immobile and mobile maize shellers [10]. These shellers save time and they reduce on the drudgery during maize shelling. However, the challenges with some of these shellers is that they are heavy [8], do not clean the maize kernels and are characterized with a broken percentage of 8.4 [1] which is

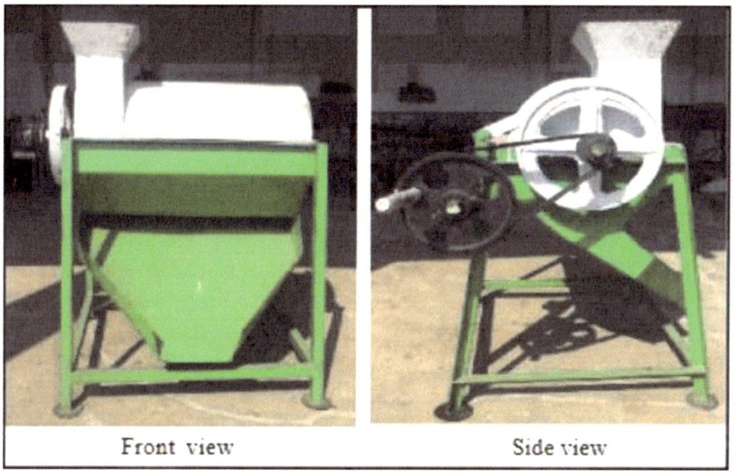

Figure 3.
Manual maize sheller [18].

Figure 4.
Motorized maize shelling [19].

above the recommended 2% [20]. Motorized maize shellers use mechanically generated power to shell the maize. To facilitate speedy shelling of maize in large scale maize production, motorized maize shellers are recommended compared to hand-operated maize sheller [2]. The output of motorized maize shellers range between 500 and 2000 kg/hour and they can be operated by tractor power take off (PTO) or engines with power varying from 5 to 15 hp depending on the equipment used [2].

3. Maize sheller design considerations

The design objective is to obtain maximum shelling performance from the equipment. The performance of shellers in terms of shelling efficiency, grain damage percentage, output capacity, cleaning efficiency, and power requirement is a function of design parameters, operating factors, physical and engineering properties of maize [21].

3.1 Design parameters

Design parameters include: cylinder diameter, cylinder speed, shelling length, clearance between the spikes and the concave, diameter holes in the concave, spike shape, size, and arrangement on the shelling drum and the blower type. Uttam et al. [11] recommended 886 rpm and 12.05–13.64% for shelling speed and moisture content, respectively [1] for the best shelling results. At these conditions, the study concluded that the shelling efficiency, cleaning efficiency, grain recovery efficiency, total grain losses, and output capacity were 87.08, 95.89, 95.48, 2.96, and 623.99 kg/h respectively. Chilur and Kumar [22] developed and evaluated the performance of a modified dehusker cum sheller. In their study, they recommended a clearance of 25 mm between the spikes and the concave for good shelling results.

3.2 Operating factors

Operating factors include grain moisture content, shelling speed, and the feeding rate. An evaluation of these factors depends on the knowledge and understanding of the equipment's mode of operation.

Shelling efficiency is increased by reducing the moisture content [23]. This can be attributed to less resistance to the removal of maize grains from the cobs due to low moisture. The grain damage percentage increases with a reduction in moisture content [1]. This can be attributed to less deformability of the grains which reduces the breakage at low moisture content. The sheller output capacity also increases with a reduction in moisture [24]. This can be attributed to the reduced time needed to remove maize grains from maize cobs as moisture content lowers. Likewise cleaning efficiency increases with a decrease in moisture content [25]. This can be attributed to the negligible moisture content of the chaff as the grain moisture content reduces.

The shelling efficiency is increased by an increase in shelling speed [23]. This can be attributed to the increased ease in the removal of maize grains from the cobs as a result of increased impacts and resistance created between the shelling drum and the concave as the shelling speed increases. Increased shelling speed increases the grain damage percentage [1]. This can be attributed to the more force exerted to the maize grains on the cobs as a result of higher cylinder speed and frequency of impacts at higher shelling speed. Increased shelling speed causes an increase in the output capacity. The output capacity of the sheller also increases with an increase in shelling speed [24]. This can be attributed to more removal of maize grains from the maize cobs due to increased impacts and resistance created between shelling drum and the concave with the increased shelling speed. Likewise, the cleaning efficiency increases with an increase in the shelling speed [25]. This may be attributed to an increase in the air flow rate produced by the sheller blower as the shelling speed raises.

Increasing the feeding rate decreases the shelling efficiency [26]. This can be attributed to the increase in unshelled grains that comes with the increase in the feeding weight as the feeding rate increases. The increased feeding weight causes an imperfect contact between concave and shelling drum. Also, increasing the feeding rate, decreases the broken grain percentage. This is due to increasing the weight entering the sheller through the hopper which acts as a cushion that reduces the effect of the grains with the shelling unit and this reduces the broken grain percentage.

To find out how different design and operating factors of maize shellers affect their performance, studies have been conducted. Aremu et al. [27] designed, constructed, and assessed the performance of the motorized maize shelling machine. The experiment used three pulleys to change the shelling speed between 623 and 886 rpm with moisture content at levels of 13, 15, and 17%. Their study noted that maize grains of lower moisture contents were easily removed from the maize cobs. This was in agreement with what [28] found out when they conducted a similar experiment under the same conditions. The study further noted that shelling speed is directly proportional to the shelling efficiency and output capacity.

In most of the earlier studies, one operation factor was studied at ago using different experiments. However, using factorial experiments, the researcher can compare all treatments that can be created by different factor levels [29]. Factorial experimentation is highly recommended because every observation gives information about all the factors in the experiment. Srison et al. [30] used a factorial experiment to study different factors affecting losses and power consumption of axial flow corn shelling unit at different levels of the main effects. The study results revealed that peg tooth clearance, concave rod clearance, and concave clearance had significant difference on the shelling losses and power consumption, but not on grain breakage. Ugwu and Omoruyi [31] conducted an experiment to find out the effect of moisture content and feeding rate on the shelling efficiency. A 2 hp electric motor was used to provide the drive through belt connections to drive the pulley on the shelling chamber. The factorial experiment was conducted using three different moisture contents and feeding rates. The feeding rates were 3.75, 4.75, and 5.75 kg/s. The moisture contents were 10, 15, and 20%. The study observed that the shelling efficiency of the maize sheller was significantly and negatively affected by moisture contents of more than 15%. The results obtained also showed that shelling efficiency of the equipment was 99.01% at a moisture content of 10%.

3.3 Physical factors

The important crop physical factors include the moisture content, the biometric properties such as length, width, arithmetic and mean diameter, shape, volume and surface area of the grains [32], grain cob ratio, grain bulky density, sphericity, angle of response, terminal velocity, one thousand grain mass, and porosity [2]. One thousand grain weight, density, sphericity, and surface area of different grains are required when designing different separating, handling, storing, and drying systems. Bulky density, true density, and porosity are needed when sizing grain hoppers and storage facilities [33]. They can also affect the rate of heat and mass transfer of moisture during aeration and drying processes. Density is used to separate materials with different densities or specific gravities.

The arithmetic mean diameter (D_a) in mm and geometric mean diameter (D_g) in mm of the grains can be calculated using Eqs. (1) and (2) according to [32].

$$D_a = \frac{(L + W + T)}{3} \tag{1}$$

$$D_g = (L \times W \times T)^{\frac{1}{3}} \tag{2}$$

where
L: length of the maize grain, mm
W: width of the maize grain, mm
T: thickness of the maize grain, mm

The sphericity (ϕ) is surface area of a sphere with the same volume of the maize grain can be determined using Eq. (3) according to [34].

$$\phi = \frac{(L \times W \times T)^{\frac{1}{3}}}{L} \tag{3}$$

The surface area, S in mm^2 of agricultural products generally indicates the patterns of behavior in a flowing fluid such as air, as well as the ease of separating additional materials from the product during cleaning by pneumatic means. The surface area of the grains can be calculated using Eq. (4) according to [33].

$$S = \pi \left(D_g\right)^2 \tag{4}$$

The bulk density of the main grains can be calculated using Eq. (5) according to [34].

$$\rho_b = \frac{4M}{\pi D^2 h} \tag{5}$$

where
ρ_b: bulky density, gcm^{-3}
M: mass of grains that fills the height of 150 cm measuring cylinder, g
D: internal diameter of glass sampler, cm
h: height of the maize in the glass jar sampler, cm
The angle of response can be calculated using Eq. (6) according to [35].

$$\theta = \tan^{-1}\left(\frac{h_0}{r}\right) \tag{6}$$

θ: angle of response, degrees
h_0: height of the maize heap, m
r: radius of the maize heap, m
For primary processing of maize, particularly maize shelling, it is important to determine these physical properties mostly dependent on moisture content. Atere et al. [36] carried out a study on the physical properties of the maize varieties commonly grown in Nigeria. Properties determined included tri-axial dimensions (length, width, and thickness), sphericity, bulky density, true density, porosity, one thousand seed grain weight, and co-efficient of static friction. The data obtained was subjected to analysis of variance (ANOVA) and least significance difference (LSD) tests. The moisture contents of maize in this experiment were 11.35, 11.34, and 11.25%. The ANOVA results showed that maize grain properties of length, thickness, and effective diameter, bulky density, true density, porosity, and response were significantly different ($p < 0.05$) within the three varieties.

3.4 Engineering factors

Engineering properties are divided into frictional and aerodynamic properties and they are used in designing equipment for solid flow, conveying systems, and separation equipment [37]. Frictional properties include the coefficient of friction and angle of response, which can be measured using the angle of response apparatus (**Figure 5**). It consists of a plywood box of 60 mm × 60 mm × 60 mm (a) and a protractor (c) for measuring the angle in degrees and provided with a fixed and adjusted plates [32]. It also has a control (b) for raising and lowering the box during

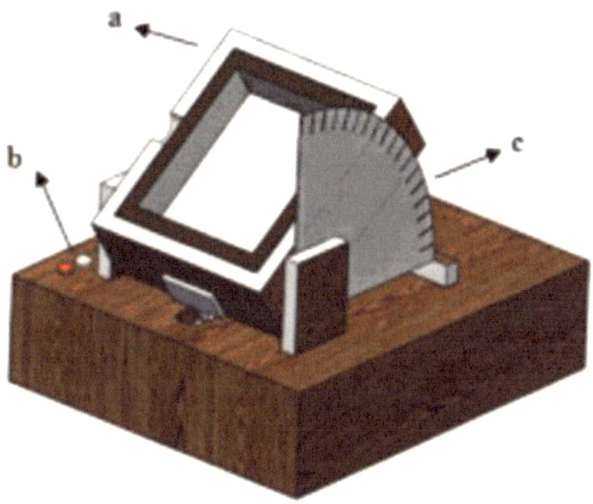

Figure 5.
Angle of response apparatus [32].

measurements. The box is filled with maize and adjustable plate inclined gradually allowing the grains to slide and assume a natural slope. The static coefficient of friction of maize grains on different surfaces can then be determined by this apparatus. Aerodynamic properties include drag coefficient and terminal velocity measured using the terminal velocity apparatus [37].

Identifying the physical and engineering characteristics of grains is important when designing, improving and optimizing the separation and cleaning equipment [34]. The engineering selection and design of grains equipment requires knowledge of these grain properties because they are of great importance in the simulation and design of these equipment. Their influence is more pronounced in problems of conceptual design where a wrong estimation of a property can lead to a design plan that is not feasible. The knowledge of maize properties also gives information about the product quality, its acceptability by different groups of consumers and its behavior in post-production, during storage, and consumption.

4. Designing a maize sheller

To ensure safe food, the equipment used for shelling maize should be designed, fabricated, and tested according to the required food grade design requirements. Mild steel can be used for maize sheller fabrication because it does not contaminate dried foods like maize grains. Besides, mild steel is smooth textured, mechanically stable, easily cleaned, and readily available at a relatively low cost. Bako and Batule [38] used mild steel to construct the shelling drum, spikes, conveyor, sieve, upper casing, hopper, exit cutes, and the frame of the maize sheller. Akoy and Ahmed [39] noted that mild steel can be used to achieve the equipment objective at the lowest cost possible. Designing a maize sheller requires designing the individual parts and then assembling them. These parts include main and other shafts, hopper, power transfer systems, and other parts.

The main shaft of the maize sheller can be designed using a hollow shaft because it has less weight, it is better in absorbing torsional loads and with great strength to weight ratio. Torsion theory [40] as shown by Eq. (7) can be used to calculate the minimum and maximum shaft diameters.

$$\frac{T}{J} = \frac{\tau}{R} \tag{7}$$

where
T: applied external torque, Nm
J: polar second moment of area of the shaft cross section
τ: shear stress at radius R and is the maximum value for both solid and hollow shafts
R: outer radius of the hollow shaft
For hollow shafts J is calculated using Eq. (8) [40].

$$J = \frac{\pi\left(D^4 - d^4\right)}{32} \tag{8}$$

where
D: outer shaft diameter
d: inner shaft diameter
Calculation of the Torque generated by the available power required to shell the maize can be done using Eq. (9) [27].

$$P = T\omega \tag{9}$$

where ω is angular velocity in rad/s calculated from Eq. (10) [27].

$$\omega = \frac{2\pi N}{60} \tag{10}$$

where N is shelling speed in rpm
Using a diameter ratio of $d = 0.833D$ and a maximum allowable shear stress $\tau_{\max}$ of 42 MNm^{-2} for a mild steel hollow shaft, the minimum shaft diameters can be calculated [40].

The concept of calculating the volume of the frustum of the pyramid using Eq. (11) can be used to size the hopper [1]. Volume of the frustum (hopper) is the difference between big pyramid volume and the small pyramid volume.

$$V = \frac{1}{3}bh \tag{11}$$

where
V: volume of the pyramid, m^3
b: base area, m^2
h: pyramid height, m
The maximum bending moment $M_{b\max}$ can be obtained by taking moments about any point along the shaft while considering all the forces acting on the shaft and their respective distances from the chosen point [1]. A shear force diagram and bending moment diagram are then drawn from which the maximum bending moment is read.

The torsional moment M_t can be determined using Eq. (12) according to [24, 27].

$$M_t = \frac{P}{2\pi N} \tag{12}$$

where
M_t: torsional moment, Nm

P: power, watts
N: speed, rpm

The bending, load, bending stress (tension and compression) can be calculated from Eq. (13) [24].

$$S_b = \frac{M_b R}{I} \quad (13)$$

But for hollow sections, $I = \frac{\pi(D^4 - d^4)}{64}$ [40].

where
S_b: bending stress, MNm^{-2}
D: outer diameter
D: internal diameter
I: moment of inertia

The torsional stress can be determined using Eq. (14) according to [41].

$$\tau_{xy} = \frac{M_t R}{J} \quad (14)$$

where
τ_{xy}: torsional stress, Nm^{-2}
M_t: torsional moment
R: outer radius of the shaft
J: polar moment of inertia
d: inner diameter of the shaft

Torsional rigidity of the shaft can be based on permissible angle of twist. The amount of twist permissible depends upon the particular application and it can vary from 0.3 m^{-1} for machine tools shaft to 3 m^{-1} for line shafting [41]. Torsional rigidity can be calculated from Eq. (15) according to [41].

$$\theta = \frac{TL}{GJ} \quad (15)$$

where
θ: angle of twist, degrees
L: length of the shaft, m
G: torsional modulus of rigidity, Nm^{-2}

The lateral rigidity of the shaft can be based upon the permissible lateral deflection for proper operation, accurate machine tool performance, shaft alignment, and other factors. The amount of deflection can be calculated by two successive integrals shown by Eq. (16) according to [40].

$$\frac{d^2y}{dx^2} = \frac{M_b}{EI} \quad (16)$$

where
M_b: bending moment, Nm^{-2}
E: modulus of elasticity, Nm^{-2}
I: moment of inertia, m^4

The sheller main shaft speed and the engine shaft speed can be related by power transfer equation shown by Eq. (17) according to [24].

$$N_1 D_1 = N_2 D_2 \quad (17)$$

where
N_1: speed of the driver pulley, rpm
D_1: diameter of driver pulley, m
N_2: speed of the driven pulley, rpm
D_2: diameter of the driven pulley, m

5. Economic feasibility of maize shelling as a business

Most fabricators, wholesalers, and retailers of maize shellers in many countries do not have definite capacity building and after-sale services to the maize sheller users [42] and no adequate instructions on equipment maintenance. Hence the entrepreneurs mostly learn on their own the operation and maintenance of their maize shellers. As a result, the economic lives of maize shellers become shorter and cause a financial loss to entrepreneurs. Thus, determining the key indicators relating to the financial feasibility of a maize shelling business is of greater importance before getting into the maize shelling business. These indicators include benefit–cost ratio and payback period [43]. The payback period is the period within which the initial investment will paid. It can be estimated using Eq. (18) according to [24].

$$P = \frac{I}{NA} \tag{18}$$

where
P: payback period, years
I: investment cost, USD
NA: net annual returns, USD

The benefit–cost ratio can be defined as the comparison of the present worth of the costs with the present worth of the benefits [42]. The benefit–cost ratio can be calculated using Eq. (19) according to [24] and it is recommended to be greater than one for the shelling business to be financially viable.

$$BC = \frac{DB}{DC} \tag{19}$$

where
BC: benefit–cost ratio
DB: discounted benefits
DC: discounted costs

$$\text{Discounted Benefits} = \sum_{t=1}^{n} \frac{B_t}{(1+r)^t}$$

$$\text{Discounted costs} = \sum_{t=1}^{n} \frac{C_t}{(1+r)^t}$$

B_t: returns for year t, USD
C_t: cost for year t, USD
t: economical life, years
r: discounted rate

6. Modification and improvement of mobile maize shellers

Modification of maize shellers can lead to improvement of the existing shellers for better performance. Most engineering designs are classified as systems created by human effort and did not exist before or improvements on the existing ones. These designs do not suddenly appear from nowhere. They result from merging technologies to meet or solve existing problems from time to time. Modification of maize shellers can be aimed at improving the performance of the existing shellers by adjusting mechanisms to certain working conditions [44]. Abagissa and Befikadu [45] noted that modification of maize shellers can result in causing no damage to maize kernels at all. Their study further revealed that the shelling efficiency was 99.67% at a moisture content of 14.7%. The evolution of motorized mobile maize shellers is a result of modification of the immobile motorized maize shellers to solve the power and transportation problems.

6.1 Case study 1: multi-purpose farm vehicle mobile maize sheller

According to [1], a study was conducted to evaluate the performance and optimize the shelling operation of the multi-purpose farm vehicle shelling technology. The study was aimed at: (i) improving the available market maize sheller and evaluate its performance and (ii) optimizing the shelling operation of the multi-purpose farm vehicle using the modified sheller. At present, transportation of maize shellers and engines (power source) from place to place is a big challenge in maize shelling. In Uganda, shellers and engines are transported on motorcycles, which not only require an extra cost, but also extra time and energy. In an effort to improve maize shelling in the country, a multi-purpose farm vehicle with a provision for hitching a maize shellers was developed to solve the power and transport problems faced by maize farmers. The three-wheeled vehicle can be used for water pumping, maize shelling, rural transport, and phone charging. This technology involves use of a multi-purpose farm vehicle power take off (PTO) power to run the maize shellers using a V-belt and a pulley. The multi-purpose farm vehicle was evaluated using a motorized market sheller and the mean broken percentage of the shelled maize was 8.43%, which was higher than the 2% recommended [20]. As a result, the holes of the concave were increased to 15 mm from 12 mm so that maize grains could easily fall through, a hollow shaft was used instead of the solid shaft for the main shelling shaft, the clearance between the concave and the spikes was modified from 22 to 25 mm which was just enough to allow the grain from being detached from the cob without damaging them and the number of the fun blades was increased from 4 to 8 [1]. The modified maize sheller was evaluated (**Figure 6**) to assess if the results were satisfactory. One way analysis of variance (ANOVA) was done using R-studio. The economic feasibility of the shelling technology was also conducted.

It was noted that the output capacity, cleaning efficiency, and grain damage percentage of the modified maize sheller was significantly different ($P < 0.05$) from the values obtained by the market maize sheller (**Table 1**). However, there was no significant difference between the shelling efficiency of the modified maize sheller and the market maize sheller. Hence in terms of shelling efficiency, both the market maize sheller and the modified maize sheller were good since their values were all above 97%.

The results of the benefit–cost analysis of the modified maize sheller powered by the multi-purpose farm vehicle are presented in **Tables 2** and **3**.

The benefit–cost-ratio and pay back period of the modified maize sheller were 1.07 and 1.37 years, respectively (**Table 3**). These results were in agreement with [42] who obtained a benefit–cost ratio of 2.34 for a maize sheller for which it

Figure 6.
Operational view of the modified multi-purpose vehicle maize sheller [1].

Performance indicator	Units	Market maize sheller	Improved maize sheller	*p*-values
Output capacity	kg/h	608.0	1581.0	$P < 0.05$
Shelling efficiency	%	97.4	98.0	$P > 0.05$
Cleaning efficiency	%	18.4	98.3	$P < 0.05$
Grain damage percentage	%	8.4	0.7	$P < 0.05$

Table 1.
Market maize sheller versus modified maize sheller [1].

Particulars	Cost, USD
Fixed cost (cost of the sheller)	577.0
Annual variable cost	2982.9
Annual gross income from shelling	3405.4
Annual net returns	422.6

Table 2.
Various costs for the modified maize sheller [1].

Particulars	Details
Payback period (years)	1.37
Benefit–cost ratio	1.07

Table 3.
Payback period and benefit–cost ratio of modified maize sheller [1].

required to be greater that one. In addition, the modified sheller investment would pay back the initial investment within 1.5 years or approximately three maize growing seasons. Hence the maize shelling operation of the modified maize sheller powered by the multi-purpose vehicle is a profitable venture for entrepreneurs.

6.2 Case study 2: two-wheel tractor mounted mobile maize sheller

According to [9], a study was conducted to develop a cost effective two wheel tractor mounted mobile maize sheller for small-scale farmers in Bangladesh in South Asia. Two-wheel tractor (power tiller) is a common tillage tool in Bangladesh agriculture because it can easily access fragmented land that is affordable to small scale farmers. Traditionally, maize shellers need to be carried from place to place by hooking with two-wheel tractor (2WT) and set it up again for shelling operation. This takes longer time for preparation of maize shelling.

To consider this problem and constraint, a small cost-effective mobile maize sheller was developed, which is mounted on the front side of the two-wheel tractor (**Figures** 7 and **8**).

So, the driver of the 2WT carry and move the sheller along in the 2WT driving position. The engine of 2WT is used as a power source for operating the maize sheller.

The mobile maize sheller eradicates the transportation problem and can start shelling operation instantly at any place since it is attached together with 2WT. It is counter clockwise rotating cylinder, axial flow type sheller and grain separated with a resistance between spike tooth and the concave. The maize sheller is attached with nuts and bolts in front of the engine base of 2WT. The operating power of the sheller comes from the fly wheel of the engine of the tractor through a V-belt and a pulley.

The shelling performance of the mobile maize sheller is shown in **Table 4**. The shelling capacity, shelling broken kernel and cylinder loss of the mobile maize sheller were 2100 kg/h, 2.3 and 0.35%, respectively. The efficiency of the mobile maize sheller was 97%.

Effective operating hours of mobile maize sheller is more than that of the traditional maize sheller (**Table 5**). This is because shelling unit of the mobile maize sheller is assembled with the transportation power unit and service providers freely carry the maize sheller to different farmers' home yards in assembly position. This therefore, reduces the maize sheller installation and starting time. The effective operating hours/day were 6.5 and 4.5 hours for the mobile maize sheller and immobile maize sheller, respectively. Mobile maize sheller saves 2 hours/day that is this sheller can be used for an additional 2 hours in day compared to the immobile maize sheller. The shelling cost for mobile maize sheller was 0.0026 USD/kg of grain which was lower than 0.012 USD/kg for the immobile maize sheller (**Table 6**).

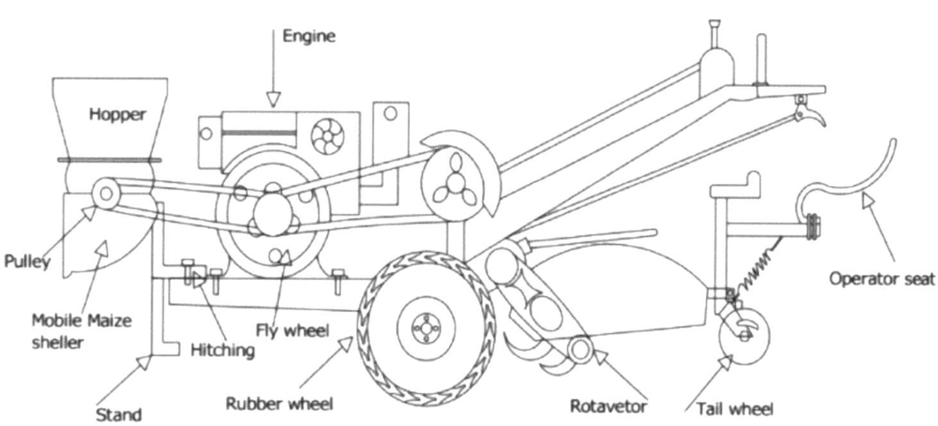

Figure 7.
Side view of the two-wheel tractor with the mobile maize sheller [9].

Figure 8.
Operational view of the two-wheel tractor mobile maize sheller [9].

Performance parameter	Units	Measured value
Cylinder speed	rpm	1250
Throughput capacity	kH/h	3150
Average shelling capacity	kg/h	2100
Cylinder loss	%	0.35
Separating loss	%	0.40
Broken kernel	%	2.20
Shelling efficiency	%	97

Table 4.
Shelling performance of the mobile maize sheller [9].

Maize sheller name	Average effective use, hours/day	Time saving, hours/day
Mobile maize sheller	6.5	2
Immobile maize sheller	4.5	—

Table 5.
Effective use hours of mobile maize sheller versus immobile maize sheller [9].

Maize sheller types	Shelling cost, USD/kg	Shelling cost, USD/year	Net return, USD/kg	Net return, USD/year	BCR
Mobile maize sheller	0.0026	3.416.72	0.012	17,646.94	5.16
Immobile maize sheller	0.012	—	—	—	—

Table 6.
Shelling cost of the mobile and immobile [9].

The lower shelling cost of the mobile sheller can be attributed to the extra two hours that it can operated per day compared to the immobile maize sheller. The benefit–cost ratio (BCR) of the mobile maize sheller was 5.15.

Figure 9.
Operational view of the Munyegera Agro-Machinery mobile maize sheller [10].

6.3 Case study 3: Munyegera Agro-Machinery mobile maize sheller

The last case study is from [10] about a mobile maize sheller (**Figure 9**) designed and fabricated by an enterprising fabricator Munyegera Agro-Machinery in Eastern Uganda with encouragement, advice, training, and initial funding from Non-Government Organization (NGO) Sasakawa 2000.

Although there is not much scientific information on its design, fabrication, and evaluation, it can be noted that this mobile maize sheller capacity is 2000–3000 kg/h [10] which is higher than most motorized immobile maize shellers. This can be attributed to the bigger shelling unit of the mobile maize sheller compared to the motorized immobile maize shellers. Operation of this mobile maize sheller requires three to four workers. Hence, whether a self-employed agent or large-scale farmer service enterprise like the Bugiri Agribusiness Initiative Development Association, youth are typically hired to operate and maintain the maize shellers which has contributed to rural enterprise growth and job creation.

Feed the Future Uganda Commodity Production and Marketing (CPM) initially cost-shared 70 these mobile maize shellers in 2015, particularly with large traders and farmer organizations linked to village agents to demonstrate the benefits of this technology [10]. On observing the benefits, some traders started buying the mobile shellers and have their village agents operate them. Apex farmer organization also purchased the mobile maize shellers to provide the mobile maize shelling service to their members. As of March 2016, many CPM clients acquired 280 mobile maize shellers [10]. CPM worked with Munyegera Agro-Machinery to train more than 200 operators in operations and maintenance, as well as maize quality control with an idea that shellers will be offering premium prices on behalf of their buyers.

7. Conclusion

This book chapter's main aim was to describe the maize shelling operations in low developing countries with focus on the need for a paradigm shift from immobile maize shellers to mobile maize shellers. Compared with immobile maize shellers, mobile maize shellers have the potential to solve the power problem as well as sheller transportation problem and the extra energy required to lift the maize shellers up and down during the shelling process. In addition, mobile maize shellers save time hence increasing their effective use hours in the field. To maximize the shelling operation, it is recommended that the moisture content of maize is maintained between 12 and 13% at a shelling speed of 880 rpm. Also, the clearance between the spikes and the concave should always be designed depending on the maximum and minimum diameters of the maize cobs.

Acknowledgements

The Presidential Initiative for Scientific Research at the School of Food Technology, Nutrition and Bioengineering, Makerere University is acknowledged for sponsoring the research and technology development leading to design and construction of the multi-purpose farm vehicle shelling unit.

Conflict of interest

The authors declare that there is no conflict of interest.

Author details

Denis Nsubuga[1*], Isa Kabenge[1], Ahamada Zziwa[1], Nicholas Kiggundu[1], Joshua Wanyama[1] and Noble Banadda[1,2†]

1 Department of Agricultural and Biosystems Engineering, Makerere University, Kampala, Uganda

2 Department of Agricultural and Biosystems Engineering, Ames, IA, USA

*Address all correspondence to: dnsubuga5@gmail.com

†Dedicated to the author Noble Banadda who passed away while this book chapter was being prepared.

References

[1] Nsubuga D, Kabenge I, Zziwa A, Kiggundu N, Banadda N. Performance evaluation and optimization of the maize shelling operation of the multi-purpose farm vehicle. Agricultural Engineering International: CIGR Journal. 2020;**22**(4):174-183

[2] Merga WD. Review on development and performance evaluation of maize sheller. International Journal of Engineering Research and Technology. 2019;**8**(5):472-481

[3] Win HH, Maung Thwin HT. Design and structural analysis of shelling cylinder for maize sheller. IRE Journal. 2019;**3**(2):389-393

[4] Teha A, Bedda T, Katema K, Aliyi I, Nur J. Pre-extension demonstration and evaluation of engine maize sheller technology in the selected AGP-II districts of Harari Region and Dire Dawa Administration. Food Science and Quality Management. 2020;**9**:14-18

[5] Hadera D. Introduction of decker type maize sheller to selected areas of eastern tigray and its socioeconomic impact assessment. Journal of Natural Science Research. 2016;**6**(1):91-95

[6] Groote HD, Kimenju SC. Consumer preferences for maize products in urban Kenya. Food and Nutrition Bulletin. 2012;**33**(2):99-110

[7] Macauley H. Cereal crops: Rice, maize, millet, sorghum, wheat. In: Proceeding of Feeding Africa Conference. Dahar, Senegal: Abdou Diof International Conference Center. 2015. Available from: https://www.afdb.org/fileadmin/uploads/afdb/Documents/Events/DakAgri2015/Cereal_Crops-_Rice__Maize__Millet__Sorghum__Wheat.pdf [Accessed: 21–23 October 2015]

[8] Okule D, Sekanyo S. User perception towards a motorized thresher (Kungula) in Uganda: A need finding survey. African Journal of Agricultural Research. 2017;**12**(12):997-1004

[9] Hossain IM, Tiwari TP, Gulandaz A, Jahan N. Development of cost effective small No-till seeder for two wheel tractor in Bangladesh. International Journal of Mechanical, Aerospace, Industrial, Mechatronic and Manufacturing Engineering. 2017;**11**(5):973-977

[10] USAID. Mobile Maize Shellers for Improved Quality and EAC Exports. 2016. Available from: https://agrilinks.org/sites/default/files/resources/ftf_proof_of_concept_technical_brief_3.pdf

[11] Uttam PN, Hardik P, Krupesh P. Design & fabrication of a motorized maize. Journal of Materials Science and Mechanical Engineering. 2018;**5**(1):5-12

[12] Wang K, Xie R, Min B, Hou P, Xue J, Li S. Review of combine harvester losses for maize and influencing factors. International Journal of Agricultural and Biological Engineering. 2012;**14**(1):1-10

[13] Benaseer S, Masilamani P, Alex AV, Govindaraj M, Selvaraju P, Bhaskaran M. Impact of harvesting and threshing methods on seed quality-A review: Agricultural Reviews. 2018;**39**(3):183-192

[14] Li Y, Tao C, Zhe Q, Kehong L, Xiaowei Y, Dandan H, et al. Development and application of mechanized maize harvesters. International Journal of Agricultural and Biological Engineering. 2016;**9**(3):15-28

[15] Špokas L, Adamčuk V, Bulgakov V, Nozdrovicḱ L. The experimental research of combine harvesters. Research in Agricultural Engineering. 2016;**62**(3):106-122

[16] Madanhire I, Chinguwa S, Ntini E, Mbohwa C. Design and simulation of

maize sheller for small scale farmers. In: Proceedings of the International Conference on Industrial Engineering and Operations Management; 23–25 October 2019; Toronto, Canada

[17] Adedipe JO, Ibiyeye DE, Onifade AO, Ekaun AA, Olatunji BT, Afolabi RT, et al. Development of a hand operated maize sheller. Australian Journal of Science and Technology. 2020;**4**(4):392-396

[18] Kumar A, Begum SH. Design, development and performance evaluation of a hand operated maize sheller. International Journal of Agricultural Engineering. 2021;**9**(10): 113-117

[19] Candia A, Saasa AR, Muzei J, Ocen P. Improving the AEATRI-motorized maize sheller to meet the market demands of commercial maize farmers. Uganda Journal of Agricultural Sciences. 2004;**9**(1):569-573

[20] SEATINI. The Status of Implementation of the East African Standard for Maize Grains and Development of a Standard for Sesame Seed in Uganda. 2014. Available from: http://www.seatiniuganda.org/publications/research/81-baseline-study-the-status-of-implementation-on-the-east-african-standard-for-maize-grains-and-development-of-a-standard-for-sesame-in-uganda-1/file.html

[21] Bhise S, Kaur A, Ramarathinam M. Moisture dependant physical prpoperties of maize (PMH-1). Acta Alimentaria. 2014;**43**(3):394-401

[22] Chilur R, Kumar S. Design and development of maize dehusker cum sheller: A technology for northern transition zone of Karnataka, India. American Journal of Engineering Research. 2014;**3**(6):127-136

[23] Oriaku E, Agulanna C, Nwannewuihe H, Onwukwe M. Design and performance evaluation of a corn de-cobbing and separating machine. American Journal of Engineering Research. 2014;**3**(6):127-136

[24] Chaudhary S. Development and performance evaluation of modified maize dehusker cum sheller [doctoral dissertation]. India: Sam Higginbottom Institute of Agriculture, Technology and Sciences; 2016

[25] Ogunlade CA, Aremu DO, Babajide NA, Akinyere AO. Design, fabrication and performance evaluation of a power (motorised) maize shelling machine. In: Proceeding of the Third International Conference on Engineering Technology Research; 5–7 August 2014; Ibadan, Nigeria

[26] Ojiako LA, Nwaogu A, Iwueke VA. Studying and evaluating the performance of locally fabricated and developed maize sheller. Journal of Biology, Agriculture and Healthcare. 2015;**1**(1):1-9

[27] Aremu DO, Adewumi IO, Ijadunola JA. Design, fabrication and performance evaluation of a motorized maize shelling machine. Journal of Biology, Agriculture and Healthcare. 2015;**5**(5):154-164

[28] Ojomo AO, Ojomo AO. Response surface methodology approach to optimising perfoemance parameters of a locally fabricated maize shelling machine. Journal of Science and Multidisciplinary Research. 2012;**4**(2): 70-79

[29] Kukreja A, Chopra P, Aggarwal A, Khanna O. Application of full factorial design for optimization of feed rate of stationary hook hopper. International Journal of Modeling and Optimization. 2011;**1**(3):2015-2209

[30] Srison W, Chuan-udom S, Saengprachatanarug K. Design factors affecting losses and power consumption

of an axial flow corn shelling unit. Songklanakarin Journal of Science and Technology. 2016;**38**(5):591-598

[31] Ugwu KC, Omoruyi A. Development and performance evaluation of maize threshing and grinding machine. American Journal of Engineering Research. 2016;**5**(10):24-29

[32] Tarighi J, Mahmoudi A, Alavi N. Some mechanical and physical properties of corn seed (var. DCC 370). African Journal of Agricultural Research. 2011;**6**(16):3691-3699

[33] Kumar BA, Rao PVK, Edukondalu L. Physical prpoerties of maize grains. International Journal of Agricultural Sciences. 2017;**9**(27):4338-4341

[34] Chhabra N, Kaur A. Studies on physical and engineering characteristics of maize, pearl millet and soybean. Journal of Pharmacognosy and Phytochemistry. 2017;**6**(6):1-5

[35] Wani IA, Sogi DS, Wani AA, Gill BS. Physical and cooking characteristics of some Indian kidney bean (*Phaseolus vulgaris* L.) cultivars. Journal of the Saudi Society of Agricultural Sciences. 2017;**16**(1):7-15

[36] Atere AO, Olalisi AP, Olukunle OJ. Physical properties of some maize varieties. Journal of Multidisciplinary Engineering Science and Technology. 2016;**3**(2):3874-3880

[37] El-Fawal YA, Tawfik MA, El-Shal AM. Study on physical and engineering properties for grains of some field crops. Misr Journal of Agricultural Engineering. 2009;**26**(4):1933-1951

[38] Bako T, Batule BJ. Design, fabrication and performance evaluation of hand operated maize sheller. Journal of American Science. 2017;**13**(8):68-76

[39] Akoy OME, Ahmed AAI. Design, construction and performance evaluation of solar cookers. Journal of Agricultural Science and Engineering. 2015;**1**(12):75-82

[40] Hearn EJ. Mechanics of Materials 1. 3rd ed. London: A Division of Reed Education and Professional Publising Ltd.; 2000

[41] Hassan BA, Abolarin MS, Olugboji OA, Ugwuoke IC. The design and construction of maize threshing machine. AU Journal of Technology. 2009;**12**(3):199-206

[42] Milufarana R, Rahman A, Alam M, Ahamed R. Economic parameter of maize sheller for custom hire service in Bangladesh. Agricultural Engineering International: CIGR Journal. 2015;**17**(2): 146-150

[43] Vishwanatha BT. An economic analysis of threshing of maize crop in Karnataka: A comparative study of mechanical versus traditional threshing methods [masters thesis]. Dharwad: Department of Agricultural Economics, University of Agricultural Sciences; 2005

[44] Wanjala FN. Design of a modified hand operated maize sheller [undergraduate thesis]. Department of Environmental and Biosystems Engineering, University of Nairobi; 2014

[45] Abagissa H, Befikadu D. Modification and testing of Jimma adjustable hand maize sheller. Journal of Multidisciplinary Science and Technology. 2015;**2**(6):1375-1377

Chapter 8

Advances and Trends in the Physicochemical Properties of Corn Starch Blends

Ulin Antobelli Basilio-Cortes, Daniel González-Mendoza, Carlos Enrique Ail-Catzim, Carlos Ceceña-Durán, Onésimo Grimaldo-Juárez, Dagoberto Durán-Hernández, Olivia Tzintzun-Camacho, Luis Antonio González-Anguiano, Ángel Manuel Suárez-Hernández, Aurelia Mendoza-Gómez, Juan Carlos Vásquez-Angulo, Adabella Suárez-Vargas, David Cervantes-García and Gabriel Luna-Sandoval

Abstract

Corn starch is one of the most widely used biopolymers in the world for various applications, due to its high production, renewable, low cost, non-toxic, biodegradable and provide great stereochemical diversity by presenting a complex structure with unique qualities that they depend on multiple factors to obtain special properties for a specific use and/or of interest. From the synthesis of the starch granule to its extraction for its subsequent use, it promotes innovative characteristics, presenting infinite functionalities applicable and/or as a substitute for synthetic polymers. However, some limitations of hydrophilicity, thermal and mechanical properties, rapid degradability and strong intra and intermolecular bonds of the polymer chains make their use difficult in the medium and long term. Enzymatic, chemical and physical methods continue to be used today, creating by-products such as polluting waste and which can be costly. Therefore, the polymeric modification of the starch granule is necessary to mitigate limitations and by-products, currently the use of starch blends is a promising trend to produce new and innovative desirable properties. This chapter describes the advances and trends in the physicochemical properties of corn starch blends *Zea mays* L. as a potential material, leader for its attractive properties and benefits that it has to offer, demonstrating that when combined with other starches from different botanical sources and/or molecular structure present unique and unequaled synergisms.

Keywords: Corn, Starch, Blends, Extrusion, Microencapsulation

1. Introduction

Corn (*Zea mays* L.) is the cereal with the highest production in the world due to the amount of nutrients that it provides beneficial for the consumer. Starch is the main component of the energy source precursor corn kernel. The main polysaccharides that make up starch are: Amylose is made up of α-D-glucose units linked by α (1,4) glucosidic bonds, while amylopectin is a branched polymer formed by linked α-D-glucose units by α (1- > 4) glucosidic bonds with branching points in the form of α (1,6) bonds. The amylose and amylopectin molecules form alternating stacks of crystalline amorphous lamellae and semicrystalline growth rings, originating highly organized granules that may differ with respect to the genotype of the botanical source of extraction, amylose is dispersed over the entire structure of amylopectin [1]. Currently corn starch is used as an emulsifying agent, encapsulant, stabilizer, colloidal gelling agent, water retention agent, among other uses due to the unique physicochemical characteristics that it presents, in addition to being accessible, non-toxic and having high yields at low cost production [2].

2. Corn starch

Corn is one of the staple foods and is used as an industrial by-product. The corn grain comes from the independent fruit called caryopsis that is inserted in the cylindrical rachis "ear", each grain or seed is limited by the number of grains per row and rows per ear. The pericarp (wall of the ovary) and testa (seed coat) join to form the wall of the ear. The corn kernel is made up of 3 main parts: "embryo", "endosperm" and "wall of the fruit". The amount of kernels produced on each ear and the number of ears that grow is generally confirmed when pollinating [3, 4]. In addition to the different derivatives of corn such as corn oil, gluten, syrup, dextrose, ethanol among others. Corn starch provides ideal characteristics for various industrial applications in the textile, food, pharmaceutical, construction fields, among others. It is composed mainly of amylose/amylopectin in different proportions and polymeric organizations. These two components of starch represent approximately ≥99% of starch by dry weight. Commonly the conformation is 75% amylopectin and 25% amylose. The polymeric structure of amylopectin provides the morphology of the granule. Amylopectin is made up of -D-glucopyranosyl chains, which are highly branched and the chains are connected to each other by 1,4 bonds, with almost 6% of 1,6 bonds forming branch points. Amylose is found in small amounts compared to amylopectin. It is an unbranched unit with 1,4-linked glucopyranosyl units, although it does not have branches, but there are some molecules that are slightly branched with 1,6-linkages. Amylopectin is the main molecule that causes the various changes on the physicochemical properties of the starch granule, changing the rheological, hygroscopic, retrogradation and leaching properties of amylose, generating new structural conformations, glass transition and maximum viscosity. On the other hand, amylose (leaching) also has a directly proportional effect on the changes that amylopectin presents, due to the response factors applied to starch [5].

Starch in the native state (without amylose/amylopectin structural modification) is available as a reserve carbohydrate in many parts of plants, including roots, tubers, cereals, and seeds. **Figure 1** presents a proposed scheme on the biosynthesis of corn starch. In general, the main enzymes for starch biosynthesis include mainly ADP-glucose pyrophosphorylases (AGPases), granule-bound starch synthases (GBSS), soluble starch synthases (SS), starch branching enzymes (BE) and starch debranching enzymes (DBE). AGPase catalyzes the first step reaction of starch biosynthesis by converting glucose 1-phosphate (Glc-1-P) and ATP to ADP-Glc and inorganic pyrophosphate (PPi) in amyloplasts. GBSS and SSS are responsible for synthesizing amylose and amylopectin, respectively. SBE introduces a branched structure by cleaving the internal chains of α-1,4-glucan and transferring the chain

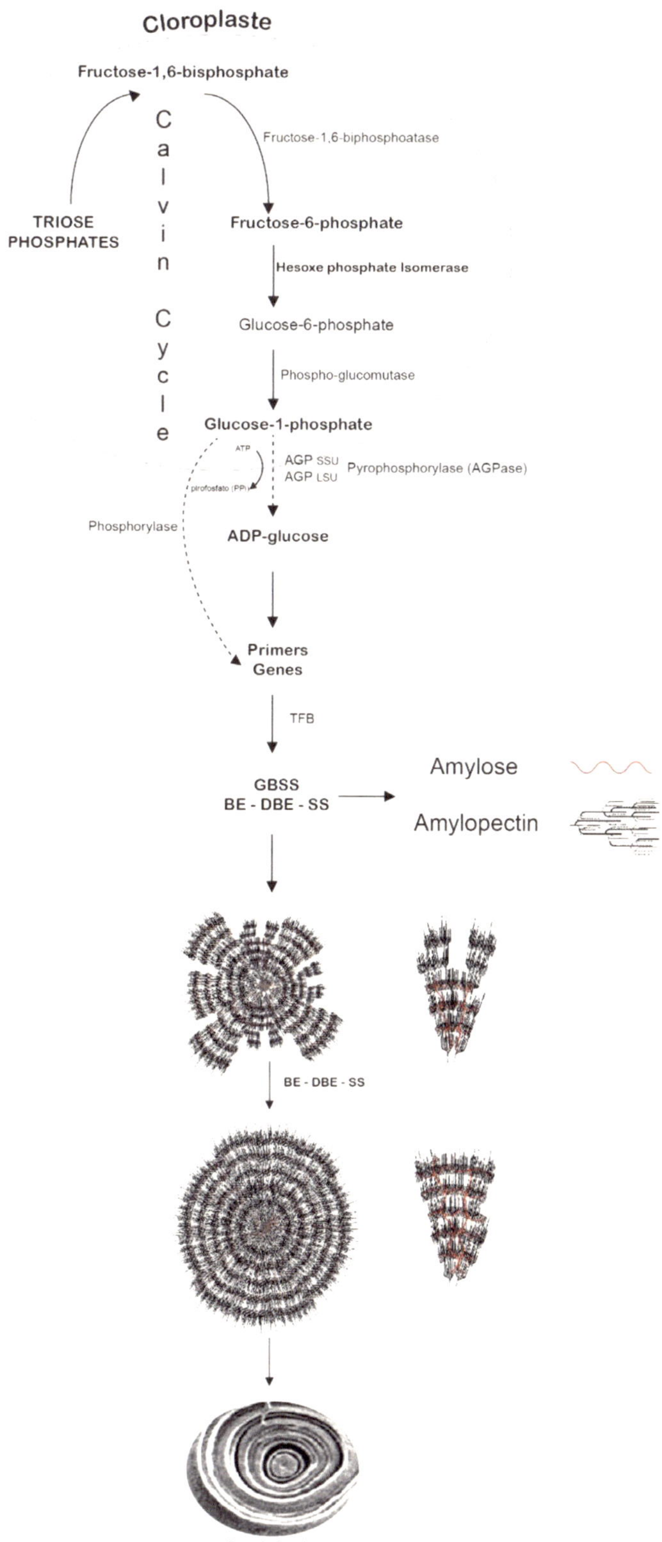

Figure 1.
Corn starch biosynthesis scheme.

segment of six or more glucose units to the C6 position of a glucosyl residue of another glucan chain. DBEs, through their α-1,6-hydrolytic activity, act on highly branched pre-amylopectin, generating polymodal distributed end chains of amylopectin. However, recent research has shown that by modifying the plant gene, variations in the content, distribution, size and polymeric organization of amylose and amylopectin are obtained [6–8]. AGPase catalyzes the first key regulatory step in the starch biosynthetic pathways present in all higher plants that produce ADP-Glc and pyrophosphate (PPi) from Glc-1-P and ATP. Plant AGPases exist as a heterotetramer (α2β2) composed of two large (LSU) and two small (SSU) subunits with slightly different molecular masses [9, 10]. SSUs are responsible for the catalytic activity of the enzyme complex, while LSU is believed to modulate enzymatic regulatory properties that increase the allosteric response of SSU to 3-phosphoglyceric acid (3-PGA) and inorganic phosphate (Pi) [9, 11]. AGPase activity is localized to both plastids and the endosperm cytosol of cereals, in contrast to other plant species where it has been reported to occur only in plastids [12]. In a previously reported subcellular fractionation experiment using corn endosperm, the highest AGPase activity was detected in the cytosol [13]. Furthermore, the genes responsible for the shrunk2 and brittle2 starch-deficient maize kernel phenotypes encode the endosperm-specific cytosolic LSU and SSU isoforms, respectively [14, 15]. This information is an indicator that plastid AGPase, by itself, is not sufficient to support normal starch biosynthesis processes in cereal endosperm, therefore, some researchers suggest that it is possible that plastid starch phosphorylase (Pho1) play an important role in the formation of primers to complete starch biosynthesis in the endosperm. Recent advances still trying to understand the functions of individual enzyme isoforms have provided new insights into how linear polymer chains (amylose) and branched bonds (amylopectin) are synthesized in cereals. Let us remember that both polysaccharides are made up of D-glucose chains linked by α (1–4) bonds. Amylose is essentially linear with α (1–4) bonds, while amylopectin is highly branched through α (1–6) bonds. Amylopectin forms type A and B polymorphic crystals that influence the arrangement of its double helices. Type A crystals produce relatively compact helices with a lower proportion of water, while type B crystals give rise to a more open structure containing a hydrated helical nucleus. X-ray diffraction studies allow us to know this type of crystal arrangements [16]. These conformations will always be different depending on the type of botanical source (TFB), as well as the enzymes involved in the formation of amylose and amylopectin.

The functional and physicochemical properties of corn starch are influenced by the amylose/amylopectin ratio, its chain length distribution and the presence of complexes in lower proportions with lipids/proteins. In its native form, corn starch has limited applications due to its low resistance to extreme processing conditions, shear, insolubility to water at room temperature, hygroscopicity among others, which are frequently found in the industry. Therefore, at present various modification techniques have been implemented that can be achieved by enzymatic, genetic, chemical, physical methods or a combination of some of these methods, which will allow a modification, mainly on the structure of amylopectin to obtain a functional starch that can overcome deficiencies and increase its usefulness for various industrial applications [17].

3. Starch blends

The use of different techniques to modify the polymeric structure of starch unfortunately have disadvantages, it can have high costs due to the use of reagents, some processes take long periods of time, low yields can generate residues that could affect the environment. Therefore, the proposals to use starch blends that promote new physicochemical characteristics that can replace conventional methods, trends that are diversifying unique properties by combining starches from different botanical sources [18]. Physical treatments are considered ecological friendly to the environment due to the absence of chemical agents and/or concentrated alkaline/acid solutions. It's essential to know the physicochemical properties of each starch to obtain the best combination in order to focus on the application, innovation or continuous improvement of some industrial type product.

3.1 Granule size

The blends of starches with different granule size and amylose content contribute to new molecular interactions between the amylose/amylopectin contents presenting significant changes in the physicochemical properties (e.g. rheological, swelling power, gel, solubility, viscosity among others), which are attributed to the content of amylose, chain length and retrogradation [18–20].

The biofilms that are formed from starch are a very promising option to avoid the excessive use of plastic (polyethylene, vinyl chloride, polystyrene and urea formaldehyde) and low and/or no deterioration on the environment. Unfortunately, with the population and industrial increase, the use of plastics continues to increase year after year, which generates millions of tons of waste after use. Polyethylene degradation is highly influenced by the biotic and abiotic environment, thus limiting effective degradation. Currently, through various investigations, the development of plastics for packaging from biodegradable materials is being promoted, using starch as the main raw material [21, 22]. The wide variety of research on the use of starches from various botanical sources (e.g. potato, corn, wheat, tapioca, rice and others) and its low cost together with comparable characteristics for film formation have shown that it is an efficient packaging raw material and a possible substitute for polyethylene, however, there are still physicochemical properties that are still under experimentation, taking into account the permeability to water vapor, mechanical properties, glass transition and hydrophobicity. The film formation process from starch granules has been described by different authors, the quality of the film is greatly affected by the amylose/amylopectin/plasticizer (glycerol) ratio, the latter being the most widely used. The content of amylose/amylopectin, a variation in the intrinsic properties of the film can be observed due to the deviation in the content of phosphorus, molar mass of starch and the biochemistry of amyloplast and chloroplast [23]. Corn is a predominant source of starch (65%) produces a film with a higher percentage of elongation, better oxygen barrier properties and a high elastic modulus, in addition [24]. The formation of biofilms from blends of cassava/corn starch with cellulose was shown to have a hydrophobic effect by reducing the affinity of starch films with water and considerably reduced the rate of permeability to water vapor, thus improving their properties. In addition, the amylose content promotes the production of a more hydrophobic film, due to the strong interactions of intermolecular bonds with glycerol [25]. New techniques for incorporating particles onto others, as reported by Farrag, et al. [26], I present reported that they prepared starch microparticles with donut-shaped morphology from two different botanical origins (type A and C for corn and pea starch granules, respectively) by means of a simple hydroalcoholic heat treatment. The donut-shaped microparticles were loaded onto films of the same botanical origin, generating greater thermal stability of the films produced. In addition, adjusting the percentage of microparticles in the thermoplastic films allowed to supply the desired amount of oxygen and water vapor to the packaged food. This is very important to keep packaged foods fresh and healthy for as long as possible. Emulsions are systems that are characterized by presenting small dispersed droplets of an immiscible liquid phase in another liquid phase, based on many applications in different industries (eg food, nutraceutical, cosmetic, hygiene, detergents, pharmaceutical and medical), however, to stabilize these emulsifying systems, synthetic surfactants (Tweens 20/60/80) or emulsifiers of animal origin (albumin/casein) are used [27]. However, they have disadvantages in their formulation because they generate foam retention due to the trapped oxygen and interactions with other suspended molecules and even the surfactant that is not compatible, an emulsion being key for the production of various food and pharmaceutical products, it is necessary to produce emulsions with solid particles of vegetable origin to stabilize the emulsions, which is called Pickering emulsions [28]. Compared to surfactant stabilized emulsions, Pickering emulsions tend to be more stable against Ostwald ripening and coalescence. Currently, starch and cellulose are used to create Pickering emulsions. Starch for its different physicochemical properties (generally recognized as safe, non-allergenic, abundant and cheap). Unfortunately, starches are still being modified to make Pickering-type emulsions, the most widely used is succinate with octenyl succinic anhydride (OSA) reagent for emulsion production. So far, the information on Pickering starch emulsions is very scattered. The links between the manufacture of Pickering emulsions and their applications are largely absent. The lack of such links seriously undermines our research efforts to better utilize emulsions for practical applications [29]. Native starch granules are not suitable for creating stable Pickering emulsions. The starch is modified to be suitable for making a Pickering emulsion. However, future trends suggest using starch blends of granule size ≤10 μm that can be compatible with starch concentration, oil and water volume, pH, ionic strength, storage conditions, processing and presence of other components for obtain a drop size (1–100 μm) [29]. Amaranth starch has shown to have sizes ≤2 μ, which is a potential candidate to produce emulsifying systems without modifying the molecular structure by chemical agents, however, physical modifications will have to be used (e.g. temperature, pH, pressure, radiation, homogenization at high revolutions among others) to obtain favorable and applicable results [30].

Grinding on native starch granules to reduce the particle size, proved to be favorable to elaborate a Pickering emulsion system [31]. The high pressure treated starch chips and ground starch particles are partially gelatinized. They can represent a mixture of rigid particle and flexible polymer model systems in emulsions. The deformability of the starch particles can be modified in situ in Pickering emulsion systems. Heating Pickering emulsions can gelatinize starch granules to different degrees. Heating can make the boundaries between adjacent particles indistinct. During heating, some amylose and amylopectin molecules leak out of the swelling granules, causing a stabilizing layer of starch granules, becoming a layer of swollen granules interpenetrated with leached starch (amylose) polymers. In this type of emulsion systems with partially gelatinized starch granules, rigid particles (granules) and flexible polymers (leached starch molecules) coexist [32, 33]. In general, the droplet size of Pickering emulsions can be adjusted using compatible starch blends, different sizes, use of physical methods, suitable starch concentrations, and processing methods.

3.2 Extrusion

Extrusion is a continuous processing method, it involves high pressure and temperature in a short time. Its main function is to mix various components. The extrusion process allows the gelatinization of the starch, the denaturation of the proteins and even the molecular degradation due to the effects of high shear depending on the screw to be used, which in turn affects the physicochemical properties of the extrudates. Many studies have been conducted to explore the relationships between extrusion processing parameters and non-food starch characteristics in order to improve processing control [34]. However, little scientific evidence has attempted to understand the relationships between the physicochemical properties of extrudates and the molecular characteristics of starch after the extrusion process using starch blends, because using starches in the native state affects the bostwick flow (viscosity and/or consistency) on the extruder barrel causing a process obstruction, therefore, previous parameters for an excellent extrusion must be considered, such as the amount of starch moisture, the temperatures of each zone of the extruder, as well as the time and type shear stress, which is crucial to achieving product quality and developing novel products [35]. The use of corn starch/cassava blends provides different degrees of gelatinization and some existing air microcells in the extrudates, attributing this effect to the extension of the puff and the fine structures of the extrudates when they were exposed to different temperatures, residence times. in the extruder and the amount of moisture present in the starch sample [36].

3.3 Encapsulation

The microencapsulation process by spray drying method allows the use of a large number of wall materials. It is essential to know the type of starch to use, it must present characteristics such as high solubility, low swelling power and viscosity, thus allowing effective encapsulation, since it can influence the properties of the emulsion, the retention of active principle, flavor and the end product shelf life. Currently there are no reports of works in which a mixture of two or more starches is implemented to be used as wall material due to the increase in viscosity, however, new trends such as using blends of porous starches, which are naturally derived from starch native by physical, chemical or biological methods. There are pores, holes and/or openings with diameters less than one micron in the structural lattices, through which molecules with smaller particle size can enter the polymeric structure and become encapsulated. The use of these possible starch blends will be an option and research topic to avoid conventional modifications on the structure of the starch, being an environmentally friendly option and without remnants of solutions with chemical reagents [37].

4. Conclusions and future trends for corn starch

Corn starch is a natural biopolymer, it has multiple physicochemical properties, an option to replace most of the synthetic polymers in the future to reduce pollution and preserve the environment. The use of blends with other starches from different botanical sources and even other biopolymers of different molecular structure (e.g. pectin, cellulose, chitosan, gelatin, alginate, hydroxyapatite, protein) are tendencies to avoid the use and generation of chemical residues that affect the environment and increase production prices. These blends are promising that demonstrate in this chapter a biocompatibility and diversity of physical properties friendly with the environment without affecting third parties. The challenge to be overcome is to completely replace the most widely used synthetic polymers around the world, through the development of biofilms based on corn starch, and how to improve the mechanical, hydrophobic and permeable properties is still being investigated. In addition, cornstarch can be used to improve filtration materials, absorbents, transport, diffusibility, hydrogels, paper, adhesives, biofuels. Therefore, the future of cornstarch blends will continue to be an encouraging proposition to generate high-value products for novel applications in various areas.

Conflict of interest

The authors declare that they have no conflict of interest.

Author details

Ulin Antobelli Basilio-Cortes[1*], Daniel González-Mendoza[1], Carlos Enrique Ail-Catzim[1], Carlos Ceceña-Durán[1], Onésimo Grimaldo-Juárez[1], Dagoberto Durán-Hernández[1], Olivia Tzintzun-Camacho[1], Luis Antonio González-Anguiano[1], Ángel Manuel Suárez-Hernández[2], Aurelia Mendoza-Gómez[2], Juan Carlos Vásquez-Angulo[2], Adabella Suárez-Vargas[3], David Cervantes-García[4] and Gabriel Luna-Sandoval[5]

1 Instituto de Ciencias Agrícolas, Universidad Autónoma de Baja California, Mexicali, Baja California, Mexico

2 Facultad de Ingeniería y Negocios, Universidad Autónoma de Baja California, San Quintín, Baja California, Mexico

3 Universidad Tecnológica de Mineral de la Reforma, Mineral de la reforma, Hidalgo, Mexico

4 Colegio de Bachilleres Guadalupe Victoria, Guadalupe Victoria, Mexicali, Baja California, Mexico

5 Universidad Estatal de Sonora, San Luis Río Colorado, Sonora, Mexico

*Address all correspondence to: antobelli.basilio@uabc.edu.mx

References

[1] Bertoft, E. (2017). Understanding starch structure: Recent progress. Agronomy, 7(3), 56. https://doi.org/10.3390/agronomy7030056.

[2] Zhu, F., & Wang, Y. J. (2013). Characterization of modified high-amylose maize starchalpha-naphthol complexes and their influence on rheological properties of wheat starch. Food Chemistry, 138(1), 256-262. https://doi.org/10.1016/j.foodchem.2012.09.097.

[3] Gopalan, C., Rama Sastri, B. V., & Balasubramanian, S. (2007). Nutritive Value of Indian foods, published by National institute of Nutrition (NIN). ICMR (Indian Council of Medical Research).

[4] Tollenaar, M., & Dwyer, L. M. (1999). Physiology of maize. In Crop yield (pp. 169-204). Springer, Berlin, Heidelberg.

[5] BeMiller, J. N. (2011). Pasting, paste, and gel properties of starch–hydrocolloid combinations. Carbohydrate polymers, 86(2), 386-423. https://doi.org/10.1016/j.carbpol.2011.05.064.

[6] Zeeman, S. C., Kossmann, J., & Smith, A. M. (2010). Starch: Its metabolism, evolution, and biotechnological modification in plants. Annual Review of Plant Biology, 61(1), 209-234. https://doi.org/10.1146/annurev-arplant-042809-112301.

[7] Jeon, J. S., Ryoo, N., Hahn, T. R., Walia, H., & Nakamura, Y. (2010). Starch biosynthesis in cereal endosperm. Plant Physiology and Biochemistry, 48(6), 383-392. https://doi.org/10.1016/j.plaphy.2010.03.006.

[8] Qu, J., Xu, S., Zhang, Z., Chen, G., Zhong, Y., Liu, L., ... Guo, D. (2018). Evolutionary, structural and expression analysis of core genes involved in starch synthesis. Scientific Reports, 8(1). https://doi.org/10.1038/s41598-018-30411-y.

[9] Tetlow, I. J., Morell, M. K., & Emes, M. J. (2004). Recent developments in understanding the regulation of starch metabolism in higher plants. Journal of experimental botany, 55(406), 2131-2145. https://doi.org/10.1093/jxb/erh248.

[10] Okita, T.W. (1992). Is there an alternative pathway for starch synthesis? Plant Physiol. 100. 560-564. https://doi:10.1104/pp.100.2.560.

[11] Ballicora, M. A., Dubay, J. R., Devillers, C. H., & Preiss, J. (2005). Resurrecting the Ancestral Enzymatic Role of a Modulatory Subunit. Journal of Biological Chemistry, 280(11), 10189-10195. https://doi:10.1074/jbc.M413540200.

[12] Hannah, L. C., & James, M. (2008). The complexities of starch biosynthesis in cereal endosperms. Current Opinion in Biotechnology, 19(2), 160-165. https://doi.org/10.1016/j.copbio.2008.02.013.

[13] Denyer, K., Dunlap, F., Thorbjornsen, T., Keeling, P., & Smith, A. M. (1996). The major form of ADP-glucose pyrophosphorylase in maize endosperm is extra-plastidial. Plant Physiology, 112(2), 779-785. https://doi:10.1104/pp.112.2.779.

[14] Patron, N. J., Greber, B., Fahy, B. F., Laurie, D. A., Parker, M. L., & Denyer, K. (2004). The lys5 mutations of barley reveal the nature and importance of plastidial ADP-Glc transporters for starch synthesis in cereal endosperm. Plant Physiology, 135(4), 2088-2097. https://doi:10.1104/pp.104.045203.

[15] Shannon, J. C., Pien, F. M., Cao, H., & Liu, K. C. (1998). Brittle-1, an adenylate

translocator, facilitates transfer of extraplastidial synthesized ADP-glucose into amyloplasts of maize endosperms. Plant Physiology, 117(4), 1235-1252. https://doi:10.1104/pp.117.4.1235.

[16] Tester, R. F., Karkalas, J., & Qi, X. (2004). Starch—composition, fine structure and architecture. Journal of cereal science, 39(2), 151-165. https://doi.org/10.1016/j.jcs.2003.12.001.

[17] Bastioli, C., Magistrali, P., & Garcia, S. G. (2013). Starch. Bio-Based Plastics: Materials and Applications, 9-33. https://doi10.1002/9781118676646.

[18] Waterschoot, J., Gomand, S. V., Fierens, E., & Delcour, J. A. (2015). Starch blends and their physicochemical properties. Starch-Stärke, 67(1-2), 1-13. https://doi10.1002/star.201300214.

[19] Waterschoot, J., Gomand, S. V., & Delcour, J. A. (2016). Impact of swelling power and granule size on pasting of blends of potato, waxy rice and maize starches. Food Hydrocolloids, 52, 69-77. http://dx.doi.org/10.1016/j.foodhyd.2015.06.012.

[20] Park, E. Y., Kim, H. N., Kim, J. Y., & Lim, S. T. (2009). Pasting properties of potato starch and waxy maize starch mixtures. Starch-Stärke, 61(6), 352-357. https://doi10.1002/star.200800029.

[21] Datta, D., & Halder, G. (2019). Effect of media on degradability, physico-mechanical and optical properties of synthesized polyolefinic and PLA film in comparison with casted potato/corn starch biofilm. Process Safety and Environmental Protection, 124, 39-62. https://doi.org/10.1016/j.psep.2019.02.002.

[22] Mali, S., Grossmann, M. V. E., & Yamashita, F. (2010). Starch films: production, properties and potential of utilization. Semina: Ciências Agrárias, 31(1), 137-156. http://dx.doi.org/10.5433/1679-0359.2010v31n1p137.

[23] Cui, C., Ji, Na., Wang, Y., Xiong, L. & Sun Q. (2021). Bioactive and intelligent starch-based films: A review. Trends in Food Science & Technology, 116 (2021) 854-869. https://doi.org/10.1016/j.tifs.2021.08.024.

[24] Luchese, C. L., Spada, J. C., & Tessaro, I. C. (2017). Starch content affects physicochemical properties of corn and cassava starch-based films. Industrial Crops and Products, 109, 619-626. https://doi.org/10.1016/j.indcrop.2017.09.020.

[25] Tavares, K. M., de Campos, A., Mitsuyuki, M. C., Luchesi, B. R., & Marconcini, J. M. (2019). Corn and cassava starch with carboxymethyl cellulose films and its mechanical and hydrophobic properties. Carbohydrate polymers, 223, 115055. https://doi.org/10.1016/j.carbpol.2019.115055.

[26] Farrag, Y., Malmir, S., Montero, B., Rico, M., Rodríguez-Llamazares, S., Barral, L., & Bouza, R. (2018). Starch edible films loaded with donut-shaped starch microparticles. LWT, 98, 62-68. https://doi.org/10.1016/j.lwt.2018.08.020.

[27] McClements, D. J., & Gumus, C. E. (2016). Natural emulsifiers — biosurfactants, phospholipids, biopolymers, and colloidal particles: Molecular and physicochemical basis of functional performance. Advances in Colloid and Interface Science, 234, 3-26. https://doi.org/10.1016/j.cis.2016.03.002.

[28] Berton-Carabin, C. C., & Schröen, K. (2015). Pickering emulsions for food applications: Background, trends, and challenges. Annual Review of Food Science and Technology, 6, 263-297. https://doi.org/10.1146/annurev-food-081114-110822.

[29] Zhu, F. (2019). Starch based Pickering emulsions: Fabrication, properties, and applications. Trends in

Food Science & Technology, 85, 129-137. https://doi.org/10.1016/j.tifs.2019.01.012.

[30] Sindhu, R., Devi, A., & Khatkar, B. S. (2021). Morphology, structure and functionality of acetylated, oxidized and heat moisture treated amaranth starches. Food Hydrocolloids, 118, 106800. https://doi.org/10.1016/j.foodhyd.2021.106800.

[31] Lu, X., Xiao, J., & Huang, Q. (2018). Pickering emulsions stabilized by media-milled starch particles. Food Research International, 105, 140-149. https://doi.org/10.1016/j.foodres.2017.11.006.

[32] Dickinson, E. (2015). Microgels—An alternative colloidal ingredient for stabilization of food emulsions. Trends in Food Science & Technology, 43, 178-188. https://doi.org/10.1016/j.tifs.2015.02.006.

[33] Sjöö, M., Emek, S. C., Hall, T., Rayner, M., & Wahlgren, M. (2015). Barrier properties of heat treated starch Pickering emulsions. Journal of Colloid and Interface Science, 450, 182-188. https://doi.org/10.1016/j.jcis.2015.03.004.

[34] Koa, S. S., Jin, X., Zhang, J., & Sopade, P. A. (2017). Extrusion of a model sorghum-barley blend: Starch digestibility and associated properties. Journal of Cereal Science, 75, 314-323. https://doi.org/10.1016/j.jcs.2017.04.007.

[35] Guo, Q. M., Joseph, M., Setia, R., Vikhona, H., Sharma, K., & Alavi, S. (2018). Extruded corn soy blends: physicochemical and molecular characterization. Journal of Cereal Science, 79, 486-493. https://doi.org/10.1016/j.jcs.2017.12.012.

[36] Seibel, W., & Hu, R. (1994). Gelatinization characteristics of a cassava/corn starch based blend during extrusion cooking employing response surface methodology. Starch-Stärke, 46(6), 217-224. https://doi.org/10.1002/star.19940460604.

[37] Wang, X., Yuan, Y., & Yue, T. (2015). The application of starch-based ingredients in flavor encapsulation. Starch Stärke, 67(3-4), 225-236. https://doi:10.1002/star.201400163.

Chapter 9

The Impact of Climate Change on Changing Pattern of Maize Diseases in Indian Subcontinent: A Review

Meena Shekhar and Nirupma Singh

Abstract

Climate change influences the occurrence, prevalence, and severity of plant pathogens. Global temperatures are predicted to rise by 2–4°C due to human activities and increased market globalization, coupled with rising temperatures, leads to a situation favorable to pest movement and establishment. Maize is an important crop after wheat and rice. Changes in rainfall distribution and temperature may result in temporary excessive soil moisture or water logging or drought in some maize producing areas leading to alterations in biotic stress factors. In Indian subcontinent warming trend in climate along the west coast, central, interior peninsula and northeast regions creates favorable conditions for diseases in maize like sorghum downy mildew (SDM) and Turcicum leaf blight (TLB). The decreasing trend of monsoon, seasonal rainfall in North India, Central India, parts of Gujarat and Kerala is suitable for post flowering stalk-rot (PFSR) which is gaining importance in maize. The outcome for any host-pathogen interaction under changing climate is not readily predictable. This review assesses the potential effects of climate change on maize pathogens and consequently on plant health. The evidence assessed indicates that climate change has already expanded pathogen's host range and geographical distribution increasing the risk of introduction of pathogens into new areas.

Keywords: climate change, maize diseases, Polysora rust, banded leaf and sheath blight, post flowering stalk rots

1. Introduction

Maize is one of the world's most widely produced and consumed cereal crops and contributes greatly to global food security. Currently 1147.7 million t is being produced jointly by over 170 countries from an area of 193.7 million ha with an average productivity of 5.75 t/ha [1]. Globally maize is being consumed as feed (61%), food (17%) and industry (22%). It has attained a position of industrial crop globally as 83% of its production in the world is used in feed, starch and bio fuel industries. Among the maize growing countries, India rank 4th in area and 7th in production, representing around 4% of world maize area and 2% of total production. In India during 2018–2019, the maize area has reached to 9.2 million ha [2].

India's production is 28.64 million t with productivity as 2.9 t/ha [3]. In India, maize is grown in two seasons, rainy (*kharif*) and winter (*rabi*). *Kharif* maize represents around 83% of maize area, while *rabi* maize correspond to 17% maize area. However, maize productivity is now threatened by global climate change [4] leading to increasing challenge by plant pathogens [5].

Climate change is affecting our agriculture due to 0.74°C average global increase in temperature in the last 100 years and atmospheric CO_2 concentration increase from 280 ppm in 1750 to 400 ppm in 2013 [6]. Throughout the twenty first century, India is projected to experience warming above the global mean. A warming trend has been observed along the west coast, in central India, the interior peninsula and Northeast India. The environment significantly, directly or indirectly, influences plants, pathogens, and their antagonists, which are strongly associated with differences in the level of losses caused by a disease, and environmental changes are often implicated in the emergence of new diseases [7]. Therefore, the changes associated with global warming may affect the incidence, severity of plant disease and influence the further coevolution of plants and their pathogens [8–12]. Plant diseases are one of the important factors which have a direct impact on global agricultural productivity and climate change will further aggravate the situation. Based on the prediction of Intergovernmental Panel on Climate Change [13], there would be an increase in 1–3°C in temperature in mid to high-latitude regions by 2050 which shows positive correlation with carbon dioxide (CO_2) concentration. The increase in CO_2 concentration and changes in rainfall pattern may have beneficial impacts on crop yields. However, moderate temperature increase (1–2°C) are likely to have negative impacts on yields of the major cereals in low-latitude regions. The regional distribution patterns of diseases getting modified as per the changes of climatic factors. On the other hand, pathogens also have the capacity to adapt to warmer conditions [14, 15]. Resistance of a disease of crop cultivars can be altered in future as per the changing situation of temperature/humidity [16, 17]. The increase in temperature and atmospheric carbon dioxide levels causes physiological changes in plants that result in increase in intensity of crop diseases. Warming may cause shifts in agro-climatic zones in which host plants migrate into new areas resulting in the emergence of new disease complexes.

A plant disease occurs only in association of a virulent pathogen, susceptible host in the presence of favorable environment [18]. A susceptible host will not be infected by a virulent pathogen if the environmental conditions are not conducive for disease, hence suitable environment is an important factor to cause disease. Therefore, the climatic condition has the potential to modify host physiology and resistance resulting into the alteration of rate and stage of pathogen. The development of plant disease is highly influenced with the environmental conditions like rainfall, relative humidity (RH), temperature and sunlight. Changes in these factors under climate change are highly likely to have an impact on the prevalence of diseases and emergence of new diseases in new area.

The impact of climate change are consistently negative for four major maize producers, together responsible for two-thirds of global maize production—United States, China, Brazil and India. The rising temperature affects flowering and leads to pests and disease build-up with significant influence on crop yield along with other parameters like soil, seed, fertilizers and agronomic practices. Maize being the third most important and widely distributed crop, can be grown in tropics, sub-tropics and temperate regions up to 50°N and S from the equator to more than 3000 meters above sea level under irrigated to semi-arid conditions.

In Indian subcontinent majority of population, depends on climate sensitive sector i.e., agriculture, forestry and fishing for livelihood and the problem of food security in our country, is very alarming and this should be addressed timely otherwise it will become more acute. Nearly one third of the population is estimated to be absolutely poor and one half of all children are malnourished in one way or another and it is going to be very difficult to ensure food security under the changing climate for the country [19]. In India 28% of the total maize produce is directly consumed as human food while 59% for poultry and animal feed, 12% for starch and dry milling and about 1% as seed. Diversified uses of maize in starch industry, corn oil, baby corns, popcorns, etc., and potential for exports has added to the demand of maize all over world. In present climate change scenario we can go for maize cultivation as being C_4 plant more suitable crop as it assimilates more CO_2 than C_3 plants. It utilizes half the quantity of water as compared to rice in *kharif* season. It is expected that the demand for maize will be double in developing world by 2050 and it will be the crop with the greatest production globally 2025 [20]. The productivity of maize is being affected adversely in present climate changing scenario, as it has direct impact on the occurrence and severity of diseases in pre and post-harvest stage in maize, which will have a serious impact on our food security. Indian subcontinent is prone to diseases like foliar diseases, ear rot and stalk rots caused by fungi and bacteria. Some economically most important diseases and major threat to the potential yield of maize are Turcicum leaf blight (TLB), Maydis leaf blight (MLB), banded leaf and sheath blight (BLSB), post flowering stalk rots (PFSR), common rust, Polysora rust, downy mildews, Pythium stalk rot (PSR) and bacterial stalk rot.

2. Present status of maize diseases in changing climatic situation

India, being a very large country, has much diversity in soils and climatic condition. Maize is grown in a wide range of environments, extending from extreme semi-arid to sub-humid and humid regions, in wet, hot climates, it has been said to thrive in cold, hot, dry or wet conditions. The impacts of climate change will vary regionally, therefore mitigation action is required immediately to limit atmospheric impact and accordingly there is a need for future studies on models which can forecast the severity of important pathogens of maize in pre-harvest and post-harvest conditions. Simultaneously, disease management strategies could be implemented in climate changing scenario with amalgamation of new strategies for sustainable food security.

In addition to climatic and atmospheric factors, the future maize productivity is dependent on climate change, as they are important for alterations in biotic stress factors [21]. Therefore, there is a risk that future maize grain yield potential might be over or underestimated if future altered effects from biotic stress factors such as diseases are ignored [22]. There is warming trend in the climate along the west coast, in central India, the interior peninsula and Northeast India [6]. However, at the regional level, increasing monsoon, seasonal rainfall has been found along the west coast, northern Andhra Pradesh and northwestern India, that creates favorable conditions for certain diseases like Polysora rust, sorghum downy mildew (SDM) and TLB, whereas excessive soil moisture or water logging is predisposing condition to bacterial stalk rot in maize. Trend of decreasing seasonal monsoon rainfall has been observed over eastern Madhya Pradesh, North India, Central India, some parts of Gujarat and Kerala such condition is suitable for post flowering stalk rot which is also gaining importance.

Disease can fluctuate according to climate variation and this relationship between day to day weather and disease development is used for disease forecasting and managing epidemics [23, 24]. Widespread occurrence of particular disease in a particular time is in climax when environmental conditions are conducive for disease development resulting in epidemics. However, increase in yield of winter maize indicates favorable changes in climate for its growth and suitability of the

S no	Disease & their distribution	Affected part	Pathogen	Favorable condition	Unfavorable condition for disease
1.	Seed and seedling blights	Seedlings	Species of *Fusarium*, *Rhizoctonia*, *Pythium Penicillium*, *Aspergillus*, *Acremonium*, *Cephalosporium* etc.	Cool (10–15°C) moist conditions, slow germination favors disease. Disease severity is affected by planting depth, soil type, seed quality, mechanical injury to seed etc.	Severity of disease reduces with the rise of temperature
2.	Turcicum leaf blight (TLB)	Foliar part at knee high stage	*Exserohilum turcicum* (Pass) Leon. & Sugs	Overwinters as mycelium and chlamydospores on debris and sporulate when cool/moderate temp. (18–27°C) coupled with high humidity.	Dry and slightly high temperature reduces disease intensity
3.	Maydis leaf blight (MLB)	Foliar part at knee high stage	*Drechslera maydis* Niskado Syn. *H. maydis*	Warm & humid conditions temp (20–32°C) and damp condition. Severe infection causes a premature death and up to 83% yield reductions.	Dry, sunny periods are unfavorable for the disease
4.	Common rust (C. rust)	Foliar part	*Puccinia sorghi* Schw	Temperate environments at tem (15–20°C) with high humidity. Uredinospore can survive at 4–30°C.	Dry weather and high temperature < 30°C is unsuitable for disease
5.	Polysora rust (P. Rust)	Foliar part	*Puccinia polysora* Underw.	Favored by high humidity at 24–28°C coupled with extended periods of dew or high humidity favors disease.	Dry weather and high temperature < 30°C is unsuitable for disease
6.	Brown stripe downy mildew (BSDM)	Foliar part	*Sclerophthora rayssiae* var. *zeae* Payak and Renfro	Favored by cool and high humid conditions coupled with dew droplets. Oospores can be viable for four years. Warm soil temperatures (28–32.5°C) favors disease.	Dry weather and high temperature < 35°C is unsuitable for disease
7.	Sorghum downy mildew (SDM)	Foliar part & in severe case tassels	*Peronosclerospora sorghi* (Weston & Uppal) Shaw	Suitable temp. for sporangia production, 17–29°C, and 21–25°C coupled with high humidity and free water droplets on leaf surface.	Dry weather and high temperature < 32°C is unsuitable for disease
8.	Rajasthan downy mildew (RDM)	Foliar part	*Peronosclerospora hetropogoni* Siradhana *et al.*	Prevails in hot and humid with temp ranges from 22.9 to 28.6°C and RH > 85.0%.	Temp. > 20.0°C and <30°C with dry weather is unfavorable for disease
9.	Brown spot	Leaf & mid rib	*Physoderma maydis* Miyake	High humidity with temp. 23–30°C and continuous rainfall is favorable for the disease.	High temp. < 30°C with less humidity is unfavorable
10	Banded leaf and sheath blight (BLSB)	Started from lower leaf & whole plant	*Rhizoctonia solani* f. sp. *Sasakii* Exner	Hot and high humidity are highly favorable condition. Suitable temperature for infection is 15–35°C.	Dry weather with high temperature is unfavorable condition

S no	Disease & their distribution	Affected part	Pathogen	Favorable condition	Unfavorable condition for disease
11.	Curvularia leaf spot	Leaf area	*Curvularia lunata*	Prolonged leaf wetness for several consecutive days. Disease occurs at temperatures, 25–35°C with 90% relative humidity.	Extreme dry weather with high temperature is unfavorable
12.	Pythium stalk rot (PSR)	Stalk at pre flowering stage	*Pythium aphanidermatum* (Eds) Fitz	High rainfall and water logged in the field with high humidity (80–100%) with high temperatures (25–35°C) favor disease severity. Free water is essential for sporangia germination.	Dry and cool weather is unfavorable for disease development
13.	Bacterial/Erwinia stalk rot (ESR)	Stalk at pre flowering stage	*Erwinia chrysanthemi* p. v. *zeae* (Sabet) Victoria, Arboleda & Munoz	High humidity, stagnant water logged condition coupled with high temperature (30°C and above).	Cool and dry weather is unfavorable for the disease
14.	Post flowering stalk rots (PFSR)	Stalk at post flowering stages	a. *Macrophomina phaseolina* b. *Fusarium moniliforme*	Warm soil temperature (30–40°C) with low moisture conditions favors disease.	Low soil temperature and wet climate is unfavorable for disease
15.	Late wilt	Stalk at post flowering stages	*Cephalosporium maydis* Samara, Sabeti & Hingorani	In India, maximum disease occurred at a constant 24°C or when the temperature varied naturally between 20 and 32°C.	Constant high temperature > 36°C is unfavorable.

Table 1.
Major disease of maize in Indian subcontinent.

region for its cultivation. Climate change is likely to lead to increase in water scarcity in the coming decades [25, 26]. If changes in atmospheric composition and global climate continue in the future as predicted, there may be chances of relocation of maize crop and relocation with occurrence of new diseases and with impact in terms of crop loss. In present scenario, the major disease of maize in Indian Subcontinent along with distribution is summarized in **Table 1**.

3. The effect of climate change on change in disease pattern scenario

Climate change modify characteristics of the pathogen, the environment, and the host, which can then drive the emergence of novel, uncommon, or adapted pathogen species, as a result there is a major shift in disease pattern during the past years as major diseases like Pythium stalk rot and bacterial stalk rots, which are gradually becoming diseases of lesser economic importance due to change in climate availability and use of sources of resistance in the development of new hybrids and varieties. However disease like Turcicum leaf blight used to occur in temperate region, now it is very common in tropics where winter maize is popular like Bihar, eastern UP, Karnataka etc., where low temperature in cropping season is favorable for the fungus. The disease of minor importance became increasingly severe and speculated epidemic proportion in coming years. In early 1960s the diseases like banded leaf and sheath blight, Polysora rust and pre harvest cob rots were considered as a disease of minor importance, however with the passes of time and changing weather condition, these diseases being arisen as major diseases not only in India but other maize growing countries of Asian region.

4. Emergence of minor diseases as major diseases due to climate change in Indian subcontinent

4.1 Banded leaf and sheath blight

Banded leaf and sheath blight (BLSB, **Figure 1**) is now become increasingly severe and economically important disease of maize during last three decades or so. During the past it was reported from Sri Lanka by [27] under the name 'Sclerotial' disease for the first time.

That time it was considered as disease of minor importance till it emerged in an epidemic form in the cooler low hills and foot hills region of Himalayas like in the district of Mandi in Himachal Pradesh. In India, this disease was recorded from Tarai region of Uttar Pradesh, for the first time in 1960 [28]. The optimum temperature for this disease is 28°C and high relative humidity (88–90%) in the first week of infection that favors rapid disease progress. However the disease development and spread becomes slow when relative humidity goes below 70% [29]. Additionally, high crop densities favors disease severity. In changing climate scenario the elevated temperature and CO_2 concentration are posing higher threat perception of BLSB and becoming severe and now it is considered as a major disease not only in Indian subcontinent but also in many part of tropical Asia wherever maize is grown in warm and humid conditions.

4.2 Polysora rust of maize

Polysora rust (**Figure 2**) also known as southern rust caused by *Puccinia polysora* is an important disease in tropical areas. It has been noticed in many parts

Figure 1.
Banded leaf and sheath blight.

Figure 2.
Polysora rust.

of the world, and it is observed in recent past from the peninsular India (coastal districts of Andhra Pradesh during winter and in Karnataka and Tamil Nadu during rainy season) on certain maize cultivars in Mysore district recorded in 1991 by [30]. The incidence of *P. polysora* has taken a heavy toll in majority of cultivars grown in Karnataka namely Mysore, Mandya, Hassan, Kolar, part of Coorg, Shimoga and Chitradurga district [31]. Disease is favored by wet/humid weather condition for infection and disease development at 12–27°C [32, 33]. Rain drizzle or even heavy dews allow spread of disease [34]. The maximum cardinal temperature for preceding infection period of *P. polysora* was estimated at 42°C and for the infection period the value ranges from 27–32°C [35–37]. In future, milder winter in temperate areas will increase diseases. Plant biomass and the concentration of CO_2 is positively correlated, however it is regulated by the other factors like availability of water and nutrients, competition between weeds, pest and pathogens. Development of biotrophic fungi such as rust and other foliar diseases on plants, promotes high concentration of carbohydrates in the host tissues. It is therefore assumed that the severity of Polysora rust and common rust will be increased in the pocket areas where is prevalence of elevated CO_2.

4.3 Post flowering stalk rot (PFSR)

The disease was first reported in 1957 from Mount Abu area of Rajasthan in India that time the disease was in traces. In recent years the severity of the disease has been increased and become an important disease, since then maize workers are quite concerned with this disease. The disease (**Figure 3**) is favored by high soil temperature (30°C-42°C) coupled with dry climatic condition. The fungi *Macrophomina phaseolina*, *Fusarium verticilloides* and *Cephalosporium* are collectively responsible for this disease. These soil-borne pathogens are favored by early spring climatic conditions when dry and slightly hot condition prevails, therefore the disease severity increases in Indian subcontinent when temperature start rising in dry weather condition. The disease is more common in hot and dry growing conditions [38]. The water stress at flowering and high soil temperature increases disease severity [39].

4.4 Curvularia leaf spot (CLS)

This is a foliar disease and caused by *Curvularia lunata,* (**Figure 4**) also common in many plant species in tropical, subtropical as well as sometimes in temperate regions. Earlier in 2000, in Indian subcontinent this disease observed in traces from most of the places. Now a days this disease (**Figure 5**) recorded in severe condition from some pockets of Rajasthan from Uttarakhand (Haridwar, Dehradun and Kashipur). Polysora rust and Curvularia leaf spot (CLS) are emerging as a potential threat in Karnataka & Rajasthan respectively. Hot and humid climate favor the development of disease during flowering to grain filling stage.

Figure 3.
Field view and symptoms of PFSR.

Figure 4.
CLS.

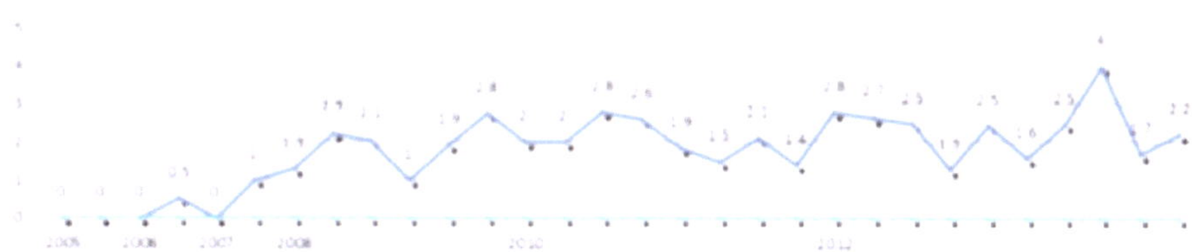

Figure 5.
Disease spectrum of Curvularia leaf spot from year 2005–2014 (source: [40]).

5. Declining of major diseases of maize with time due to climate change

5.1 Pythium stalk rots

During the past, in early 1980s Pythium stalk rot (**Figure 6**) was in severe condition and use to cause extensive damage to the crop in the lowlands of Indian subcontinent. This disease caused by *Pythium aphanidermatum*, temperature and relative humidity are the main factors to determine severity of the disease. Most favorable condition for the disease development is high temperature range 30–35°C with relative humidity of 80–100%. In past decades the disease was considered as major disease, now over the year passed the incidence/occurrence of the disease reduced [40] due to altered precipitation, various changes in this pathogens may occurred

Figure 6.
Pythium stalk rot.

which in general include coincidence of pathogen lifecycle events with host plant stages and/or natural antagonists/synergists. Therefore regional precipitation and distribution patterns led to unfavorable condition for development of sporulation for secondary infection. However, projected warmer and drier summers may hinder most fungal diseases, and finally slow down or completely inhibit disease progress of other foliar diseases. As a result the regional distribution patterns of this diseases may be modified and stated to infect *rabi*/winter maize due warming effect.

6. Potential climate change effects on ear rots and related mycotoxin contamination

Mycotoxins contamination in pre and post-harvest maize is most important and the attention focused this important disease in terms of their presence and toxicity, including variety of toxins viz., aflatoxins, ochratoxins, fumonisins, trichothecenes, and zearalenone, are secondary metabolites produced by various fungi. *Aspergillus flavus* and produces aflatoxin, *Fusarium verticillioides* produces fumonisin, and produces deoxynivelanol and zearalenone [41, 42]. Among these toxins, aflatoxin is the most important in terms of health risk and high toxicity produced by *A. flavus* and by *A. parasiticus,* which are most prevalent in tropical and subtropical regions where high temperature and drought conditions prevailed. Aflatoxins are highly carcinogenic, produced by *A. flavus* and *A. parasiticus* [43].

Aflatoxins are a group of 20 related fungal metabolites, the major ones are aflatoxin B_1 B_2 G_1 and G_2. Among them AFB_1 is the most potent naturally occurring liver carcinogen [44]. Ingestion of aflatoxins in contaminated food or feed results in aflatoxicosis, while long-term exposure of moderate to low concentration of causes chronic toxicity and immune system disorders [45, 46]. Aflatoxin contributes to significant economic losses in maize which prevents commodities from meeting internation al standards governing agricultural trade and food safety. High temperatures and low rainfall favors the production of *A. flavus* conidia and their dispersal.

Several studies reported that the high soil temperature and drought stress are positively correlated with aflatoxin contamination and increased incidence of aflatoxigenic strains or species [47–49]. The high temperature coupled with dry weather conditions favors growth, conidia formation and dispersal of fungi *A. flavus*. These factors are contributing to high concentrations of aflatoxins. Therefore these environmental factors has an important role in development of aflatoxin in maize [50–55]. High temperatures and low rainfall also favor production of conidia of *A. flavus* and their dispersal. The most important environmental influences on fumonisin risk are insect damage to grain and moisture stress in maize plants [56, 57].

The Indian subcontinent having land of diversity with diversified climatic condition. If hot humid weather prevailed at the time of critical stage of maize crop like North east, some part of western hills with limited sunshine hours at the maturity of the maize crop are predisposed condition to mycotoxin contamination. Sometimes due to unpredictable weather condition the *rabi* and *kharif* maize crop faces the same situation and ultimately spoiled due to mycotoxin contamination. Extreme dry condition at the time of flowering is also one of the predisposing factors to AFB_1 contamination. In Indian subcontinent there is extreme diversity in soils and climatic condition and maize is grown in a wide range of environments, extending from extreme semi-arid to sub-humid and humid regions. Mold inoculums that occur naturally throughout the environment in all over the world. Although maize is grown mainly in wet, hot climates, it has been said to thrive in cold, hot, dry or wet conditions.

Hence, in future there are more chances of mycotoxin contamination due to rise in temperature and unpredictable rainfall in Indian subcontinent. Maize crop from tropical and/or sub-tropical areas are affected more frequently and severely by aflatoxin contamination, but temperate areas could be of increasing importance due to climate change. In the near future, there is reason to believe that increased climate variability associated with climate change trends may result in higher pre-harvest levels of mycotoxins in Indian subcontinent, posing both economic and health risks for maize crop and food security. The occurrence of aflatoxin in maize is strongly influenced by weather during and after the growing season. Cool, wet growing seasons may delay grain maturity in maize, and result in mold contamination in the field such cobs/grains are prone to develop aflatoxin contamination. Climate change is likely to lead to an increase in hot and dry spells; this implies an expectation of increased risk of aflatoxin contamination. Insect damage of grain and moisture stress in maize favor fumonisin contamination of maize [56, 57]. The optimal conditions for fumonisin production are a temperature close to 30°C and high water activity [34, 58]. It is therefore important to determine impacts of climate change to future food security, in terms of mycotoxin-related economic and health risks. If current climate patterns continue in this century, aflatoxin and fumonisin concentrations in maize will likely increase, more aggressive isolates of *F. graminearum* occur. The overall effect may likely be increased economic and health risks, particularly due to increased aflatoxin concentrations in maize.

7. Critical analysis for future impact

7.1 Effect of CO_2 concentration

The concentration of CO_2 is positively correlate with plant biomass. However the process is regulated by the factors like availability of water and nutrients, competition between weeds, pest and pathogens. When plant is infected with biotrophic fungi such as rust and other foliar diseases, at that time the concentration of carbohydrates in the host tissues is increased. It is therefore can be assumed that the high concentration of carbohydrates in the host tissue promotes the development of biotrophic fungi such as rust [59]. Hence, there may be chances of increase in severity of Turcicum leaf blight, Polysora rust as well as common rust in the pockets where elevated CO_2 is prevailed.

7.2 Effect of temperature

Changes in temperature may favors the development of dormant pathogens, warmer mean air temperatures in India in early springs especially during winter, favors post flowering stalk rots in winter maize. However, warmer and drier summers may hinder spread of most fungal pathogens, and finally slow down or completely inhibit disease progress of foliar diseases like Turcicum leaf blight, Maydis leaf blight, banded leaf and sheath blight and downy mildews etc., resulting in the regional distribution patterns of these diseases is going to be modified and foliar diseases started to infect *rabi*/winter maize due to warming effect. Pathogens also have the capacity to adopt warmer conditions [14, 15] and temperature/humidity dependent on disease resistance of crop cultivars may be altered in the future [16, 17]. Therefore TLB prevalent cooler and temperate region, but now common in tropical in wither maize. Stalk rot diseases are anticipated to increase in hot and dry areas in summer crops as well as in winter crops of Bihar, Rajasthan, Andhra Pradesh and Karnataka due to rise in temperature in early springs at the time of flowering. Increasing trend in cultivation of *rabi*/winter maize in various states as a result host is available throughout the year round in major maize growing areas leads to multiplication of inoculum of soil borne disease which is an important factor for increasing the extent of crop losses. Therefore, the total number of diseases may not change dramatically, but there might be some changes in relocation and diversification of maize diseases in future range in India [60] also considered stalk rots of maize, particularly related to sweet corn under temperate climatic conditions. The importance of *Fusarium*/*Gibberella* stalk rots may increase in sweet corn, whereas the future importance of Pythium stalk rot might decrease.

7.3 Impact on ear/cob rots

Ear rots is associated with mycotoxin contamination with negative human and animal health consequences truly an alarming issue. The potential risk may be expected to increase in a future climate change scenario. If, temperature, drought, and insect injury in subtropical and tropical regions would increase, an increase the incidence of *Aspergillus flavus* and *F. verticillioides* may occur, consequently mycotoxin contamination will increase. If hot humid weather prevailed at the time of critical silking stage of maize crop in North east, some part of Western hills with limited sunshine hours at the maturity of the maize crop are suitable condition to mycotoxin contamination. Extreme dry condition at the time of flowering is also one of the predisposing factors to AFB_1 contamination. Although maize is grown mainly in wet, hot climates, it has been said to thrive in cold, hot, dry or wet conditions. Hence, in future there are more chances of mycotoxin contamination due to rise in temperature and unpredictable rainfall in Indian subcontinent.

8. Suggestions for new management strategies

In present climate changing scenario, the choice of crop management practices should be based on the prevailing situation. It is therefore important that weather-based disease monitoring, inoculums monitoring, should be done time to time. There is an urgent need to develop maize disease forecast module.

8.1 Rescheduling of crop planting

The changes in global climate changes have direct impact on temperature and rainfall pattern may result in shrinking of crop growing seasons with extreme problems of diversification and relocation of maize diseases. Therefore, the rescheduling crop planting dates as per suitability to maize crop in changing environment scenario is needed urgently. Maize disease management strategies needs to be changed in accordance with the projected changes in disease incidence and extent of crop losses in view of the changing climate.

8.2 Sensitization of farmers/stakeholders

In view of the impacts of future climate change on sustainability and productivity of maize, there is an urgent need to sensitize the stakeholders, farmers/growers, extension workers, about the diversification of major diseases at zonal and regional level and management strategies to cope with the situation. Sensitization of farmers, stakeholders, industries and exporters about the importance of mycotoxin/aflatoxin and their management strategies is also needed. This can be achieved through organization of awareness campaigns, training and capacity-building programmes, development of learning material and support guides for different risk scenarios.

8.3 Breeding climate-resilient varieties

In order to minimize the impacts of climate and other environmental changes, it will be crucial to breed new varieties for improved resistance to abiotic and biotic stresses such as resistant to cold and heat stress, as well as for drought and water log condition. Considering erratic monsoon late onset and/or shorter duration of winter, there is chance of delaying and shortening the growing seasons for certain *rabi*/cold season crops.

8.4 Screening of pesticides with novel mode of actions

There is a need of screening of nano-pesticides ingredient against important maize diseases. The application of zinc oxide nano particles against disease powdery mildews has antifungal activity [61, 62]. The salicylic acid is associated with plant defense responses which enhance plant vigor and abiotic stress tolerance, independent of their insecticidal action [63–67]. This gives an insight into investigating role of insecticides in enhancing stress tolerance in plants.

9. Conclusion

As the climate continues to warm in response to further greenhouse gas emissions, high temperature extremes will become hotter and cold extremes will become less cold. The amount of future global warming is closely related to cumulative CO_2 emissions that weakens our ecosystems and may support pest and disease dispersal and incidence. It has impact on plant physiology and structure, as a result vulnerability of plants towards pests and diseases may increased. Degrading ecosystems and water scarcity can affect food security and livelihoods and contribute to economic crises, forced migration and conflicts, pest and disease risks also. Indian subcontinent, being tropical area, is more challenged with impacts of impeding changes. Dealing with the climate change which is a tedious task due to its complexity, unpredictability, uncertainty and differential impacts over time period and places. However impact of climate change on crop production mediated through changes in populations of insect-pests need to be given careful attention for planning and devising adaptation and mitigation strategies for future pest management programmes. There is a need to combine both durable multiple-disease and multiple-insect resistance, using gene transfer and genome editing technology would greatly help against the pathogens and insect pests where no native resistances are available in elite breeding materials. Also steps to be taken to increase our adaptive capacity urgently so that the support to adaptation research, developing regionally differentiated contingency plans for temperature and rainfall related risks and also seasonal weather forecasts and their applications for reducing risks can be taken care. Evolvement of new land use systems, including heat and drought tolerant varieties, adapted to climatic variability and climate change. By implementing international standards for phytosanitary measures, may help countries to prevent the introduction and spread of harmful pests and to preserve biodiversity. Preserving biodiversity may helps to improve plant resilience and mitigate the impact of climate change on plant health.

Author details

Meena Shekhar[1*] and Nirupma Singh[2]

1 Division of Plant Quarantine, ICAR-NBPGR, New Delhi, India

2 Division of Genetics, ICAR-IARI, New Delhi, India

*Address all correspondence to: shekhar.meena@gmail.com

References

[1] FAO. World Food and Agriculture—Statistical Yearbook 2020. Rome: FAO; 2020

[2] Department of Agriculture and Cooperation. 2020. Available from: http://dacnet.nic.in

[3] ICAR-IIMR Annual Report 2020. Ludhiana-141004: ICAR-Indian Institute of Maize Research, Punjab Agricultural University Campus

[4] Porter BA. Global climate change and coastal tourism: Recognizing problems, managing solutions and future expectations. Journal of Ecotourism. 2019;**18**(2):195-197. DOI: 10.1080/14724049.2018.1459069

[5] Mueller DS, Wise KA, Sisson AJ, Allen TW, Bergstrom GC, Bissonnette KM, et al. Corn yield loss estimates due to diseases in the United States and Ontario, Canada, from 2016 to 2019. Plant Health Progress. 2020;**21**: 238-247

[6] Gautam HR, Bhardwaj ML, Kumar R. Climate change and its impact on plant diseases. Current Science. 2013;**105**(12): 1685-1691

[7] Anderson PK, Cunningham AA, Patel NG, Morales FJ, Epstein PR, Daszak P. Emerging infectious diseases of plants: Pathogen pollution, climate change and agro-technology drivers. Trends in Ecology & Evolution. 2004;**19**:535-544

[8] Chakraborty S. Potential impact of climate change on plant-pathogen interactions. Australasian Plant Pathology. 2005;**34**:443-448

[9] Burdon JJ, Thrall PH, Ericson AL. The current and future dynamics of disease in plant communities. Annual Review of Phytopathology. 2006;**44**:19-39

[10] Garrett KA, Dendy SP, Frank EE, Rouse MN, Travers SE. Climate change effects on plant disease: Genomes to ecosystems. Annual Review of Phytopathology. 2006;**44**:489-509

[11] Crowl TA, Crist TO, Parmenter RR, Belovsky G, Lugo AE. The spread of invasive species and infectious disease as drivers of ecosystem change. Frontiers in Ecology and the Environment. 2008;**6**:238-246

[12] Eastburn DM, McElrone AJ, Bilgin DD. Influence of atmospheric and climatic change on plant–pathogen interactions. Plant Pathology. 2011;**60**: 54-69

[13] IPCC. Climate change 2007: Synthesis report. In: Pachauri RK, Reisinger A, editors. Contribution of Working Group I, II and III to the Fourth Assessment Report of the Intergovernmental Panel on Climate Change (IPCC). Geneva, Switzerland: IPCC; 2007

[14] Zhan J, McDonald BA. Thermal adaptation in the fungal pathogen *Mycosphaerella graminicola*. Molecular Ecology. 2011;**20**(8):1689-1701. DOI: 10.1111/j.1365-294X.2011.05023

[15] Mboup M, Fischer I, Lainer H, Stephan W. Trans-species polymorphism and allele-specific expression in the CBF gene family of wild tomatoes. Molecular Biology and Evolution. 2012;**29**:3641-3652

[16] Huang YJ, Evans N, Li ZQ, Eckert M, Chevre AM, Renard M, et al. Temperature and leaf wetness duration affect phenotypic expression of Rlm6-mediated resistance to *Leptosphaeria maculans* in Brassica napus. The New Phytologist. 2006;**170**:129-141

[17] Juroszek P, von Tiedemann A. Potential strategies and future requirements for plant disease management under a changing climate. Plant Pathology. 2011;**60**:100-112

[18] Legreve A, Duveiller E. Preventing potential diseases and pest epidemics under a changing climate. In: Reynolds MP, editor. Climate Change and Crop Production. Wallingford, UK: CABI Publishing; 2010. pp. 50-70

[19] Dev SM, Sharma AN. Food Security. Food Security in India: Performance, Challenges and Policies. Oxfam India Working Paper Series VII. September 2010

[20] Rosegrant MW, Zhu T, Msangi S, Sulser TB. The impact of biofuels on world cereal prices. In: Background Brief in Support of Testimony to United States Congressional Briefing on the World Food Situation. Washington, DC: IFPRI; 2008

[21] Chakraborty S, Tiedemann AV, Teng PS. Climate change: Potential impact on plant diseases. Environmental Pollution. 2000;**108**:317-326

[22] Boonekamp PM. Are plant diseases too much ignored in the climate change debate? European Journal of Plant Pathology. 2012;**133**:291-294. DOI: 10.1007/s10658-011-9934-8

[23] Coakley SM. Climatic Variability in the Pacific Northwest and its effect on stripe rust disease of winter wheat. Climate Change. 1979,**2**:33-51

[24] Scherm H, Yang XB. Interannual variations in wheat rust development in China and the United States in relation to the El NinÃo/southern oscillation. Phytopathology. 1995;**85**

[25] Lobell D, Burke M, Tebaldi C, Mastrandera M, Falcon W, Naylor R. Prioritizing climate change adaptation needs for food security in 2030. Science. 2008;**319**:607-610

[26] Hendrix CS, Glaser SM. Trends and triggers: Climate, climate change and civil conflict in Sub Saharan Africa. Political Geography. 2007;**26**:695-715

[27] Bertus L. A sclerotial disease of maize (*Zea mays* L.) due to *Rhizoctonia solani* Kuhn. In: Year Book. Ceylon: Deptt. Agric.; 1927. pp. 46-48

[28] Payak MM, Renfro BL. Diseases of maize new to India. Indian Phytopathalogy. 1966;**3**:14-18

[29] Sharma RC, Rai SN, Batsa BK. Identifying resistance to banded leaf and sheath blight of maize. Indian Phytopathology. 2005;**58**(1):121-122

[30] Payak MM. Introduction of *Puccinia polysora,* polysora rust of maize in India. Current Science. 1994;**66**(4):317-318

[31] Anonymous. Forty-sixth Ann. Prog. Rep. of *Kharif* 2003. New Delhi: All India Coordinated Maize Improvement Project, IARI; 2003. pp. 1-62

[32] Hollier CA, King SB. Effect of dew period and temperature on infection of seedling maize plants by *Puccinia polysora*. Plant Disease. 1985;**69**:219-220

[33] Melching JS. Corn rusts: Types, races and destructive potential. Proc. 30th Annu Corn Sorghum Conf. 19759:90-115

[34] Reid LM, Nicol RW, Ouellet T, Savard M, Miller JD, Young JC, et al. Interaction of *Fusarium graminearum* and *F. moniliforme* in maize ears: disease progress, fungal biomass, and mycotoxin accumulation. Phytopathology. 1999;**89**:1028-1037

[35] Cammack RH. Studies on *Puccinia polysora* underw. III. Description and life cycle of *P. polysora* in West Africa. Transactions of the British Mycological Society. 1959;**42**:55-58

[36] Godoy D, Randle G, Simpson AJ, Aanensen DM, Pitt TL, Kinoshita R, et al. Multilocus sequence typing and evolutionary relationships among the causative agents of melioidosis and glanders, *Burkholderia pseudomallei*

and *Burkholderia mallei*. Journal of Clinical Microbiology. 2003 May;**41**(5):2068-2079. DOI: 10.1128/JCM.41.5.2068-2079.2003. Erratum in: Journal of Clinical Microbiology. 2003;**41**(10):4913

[37] Pivonia S, Yang XB. Relating epidemic progress from a general disease model to seasonal appearance time of rusts in the United States: Implications for soybean rust. Phytopathology. 2006;**96**:400-407

[38] Doohan FM, Brennan J, Cooke BM. Influence of climatic factors on *Fusarium* species pathogenic to cereals. European Journal of Plant Pathology. 2003;**109**:755-768

[39] Smith E, McLaren M. Effect of water stress on colonization of maize roots by root infecting fungi. African Plant Protection. 1997;**3**:47-51

[40] ICAR-IIMR. Annual Report 2014-15. ICAR-Indian Institute of Maize Research; 2015

[41] Cardwell KF, Desjardins A, Henry SH, et al. Mycotoxins: The Cost of Achieving Food Security and Food Quality. St. Paul, MN: American Phytopathological Society; 2001

[42] Miller JD. Mycotoxins in small grains and maize: Old problems, new challenges. Food Additives & Contaminants. Part A, Chemistry, Analysis, Control, Exposure & Risk Assessment. 2008;**25**(2):219-230. DOI: 10.1080/02652030701744520

[43] Bennett JW, Christensen SB. New perspectives on aflatoxin biosynthesis. Advances in Applied Microbiology. 1983;**29**:53-92

[44] IARC. IARC monographs on the evaluation of carcinogenic risks to humans. In: Some Naturally Occurring Substances: Food Items and Constituents, Heterocyclic Aromatic Amines and Mycotoxins. Vol. 56. Lyon: IARCPress; 1993. pp. 245-395

[45] Gong YY, Egal S, Hounsa A, Turner PC, Hall AJ, Cardwell KF, et al. Determinants of aflatoxin exposure in young children from Benin and Togo, West Africa: The critical role of weaning. International Journal of Epidemiology. 2003;**32**:556-562

[46] Turner PC, Moore SE, Hall AJ, Prentice AM, Wild CP. Modification of immune function through exposure to dietary aflatoxin in gambian children. Environmental Health Perspectives. 2003;**111**:217-220. DOI: 10.1289/ehp.5753

[47] Horn BW, Dorner JW. Regional differences in production of aflatoxin B1 and cyclopiazonic acid by soil isolates of *Aspergillus flavus* along a transect within the United States. Applied and Environmental Microbiology. 1999;**65**: 1444-1449

[48] Jaime-Garcia R, Cotty PJ. Crop rotation and soil temperature influence the community structure of *Aspergillus flavus* in soil. Soil Biology and Biochemistry. 2010;**42**:1842-1847. DOI: 10.1016/j.soilbio.2010.06.025

[49] Orum TV, Bigelow DM, Cotty PJ, Nelson MR. Using predictions based on geostatistics to monitor trends in *Aspergillus flavus* strain composition. Phytopathology. 1999;**89**:761-769

[50] Cotty PJ, Jaime-Garcia R. Influences of climate on aflatoxin producing fungi and aflatoxin contamination. International Journal of Food Microbiology. 2007;**119**:109-115. DOI: 10.1016/j.ijfoodmicro.2007.07.060

[51] Diener U, Cole R, Sanders T, Payne G, Lee L, Klich M. Epidemiology of aflatoxin formation by *Aspergillus flavus*. Annual Review of Phytopathology. 1987;**25**:249-270

[52] McMillian W, Wilson D, Widstrom N. Aflatoxin contamination of preharvest corn in Georgia: A six-year study of insect damage and visible *Aspergillus flavus*. Journal of Environmental Quality. 1985;**14**:200-202

[53] Payne GA, Cassel DK, Adkins CR. Reduction of aflatoxin levels in maize due to irrigation and tillage. Phytopathology. 1985;**75**:1283-1283

[54] Payne G, Widstrom N. Aflatoxin in maize. Critical Reviews in Plant Sciences. 1992;**10**:423-440

[55] Payne GA. Ear and kernel rots. In: White DG, editor. Compendium of Corn Diseases. St. Paul, MN, USA: The American Phytopathology Society Press; 1999. pp. 44-47

[56] Miller JD. Factors that affect the occurrence of fumonisin. Environmental Health Perspectives. 2001;**109**(Suppl. 2):321-324

[57] Munkvold GP. Cultural and genetic approaches to managing mycotoxins in maize. Annual Review of Phytopathology. 2003;**41**:99-116

[58] Marin S, Magan N, Serra J, Ramos AJ, Canela R, Sanchis V. Fumonisin B1 production and growth of *Fusarium moniliforme* and *Fusarium proliferatum* on maize, wheat, and barley grain. Journal of Food Protection. 1999;**64**:921-924

[59] Chakraborty S, Murray G, White N. Potential impact of climate change on plant diseases of economic significance to Australia. Australasian Plant Pathology. 2002;**27**:15-35

[60] Boland GJ, Melzer MS, Hopkin A, Higgins V, Nassuth A. Climate change and plant diseases in Ontario. Canadian Journal of Plant Pathology. 2004;**26**: 335-350

[61] Lamsal K, Kim SW, Jung JH, Kim YS, Kim KS, Lee YS. Inhibition effects of silver nanoparticles against powdery mildews on cucumber and pumpkin. Mycobiology. 2011, 2011; **39**(1):26-32

[62] He L, Liu Y, Mustapha A, Lin M. Antifungal activity of zinc oxide nanoparticles against *Botrytis cinerea* and *Penicillium expansum*. Microbiological Research. 2011;**166**:207-215

[63] Gonias ED, Oosterhuis DM, Bibi AC, Brown, RS. Yield, growth and physiology of Trimax TM treated cotton. Summaries of Arkansas Cotton Research. 2003. pp. 139-144

[64] Theilert WA. Unique product: The story of the imidacloprid stress shield. Pflanzenschut Nachrichten Science Forum Bayer. 2006;**59**:73-86

[65] Horii A, McCue P, Shetty K. Enhancement of seed vigour following insecticide and phenolic elicitor treatment. Bioresource Technology. 2007;**98**:623-632

[66] Chiriboga A, Herms DA, Royalty RN. Effects of Imidacloprid (Merit®2F) on Physiology of Woody Plants and Performance of Two Spotted Spider Mite, Fall Webworm, and Imported Willow Leaf Beetle. Bayer Environmental Science; 2009

[67] Ford KA, Casida JE, Chandranb D. Neonicotinoid insecticides induce salicylate associated plant defense responses. Proceedings of the National Academy of Sciences of the United States of America. 2010;**107**:17527-17532